T0392771

GREEN AUTOMATION

Increasing Sustainability, From Industry to Our Home

Francisco José Gomes da Silva
Department of Mechanical Engineering
ISEP - School of Engineering, Polytechnic of Porto
Porto, Portugal

CRC Press is an imprint of the
Taylor & Francis Group, an **informa** business
A SCIENCE PUBLISHERS BOOK

Cover image: Courtesy of Mr. Mário Cardoso.

First edition published 2024
by CRC Press
2385 NW Executive Center Drive, Suite 320, Boca Raton FL 33431

and by CRC Press
4 Park Square, Milton Park, Abingdon, Oxon, OX14 4RN

CRC Press is an imprint of Taylor & Francis Group, LLC

Library of Congress Cataloging-in-Publication Data (applied for)

ISBN: 978-1-032-23489-2 (hbk)
ISBN: 978-1-032-23490-8 (pbk)
ISBN: 978-1-003-27793-4 (ebk)

DOI: 10.1201/9781003277934

Typeset in Times New Roman
by Radiant Productions

Preface

Our life is conditioned by the environment. The environment is largely conditioned by the way we live. The extreme atmospheric phenomena that we observe more and more frequently clearly shows that we do not know how to respect the environment. We should have reversed our behavior by now. As we have not yet done it in a concerted way, it is time to do it vigorously, immediately.

The attitude that needs to be taken should come from consumers, forcing companies to change their attitude. This involves respecting nature and reversing the harm caused to date. The damage caused to nature comes from the most diverse sectors, from agriculture and farming to services. However, the industry assumes special relevance. It is precisely in industry that automation could be most useful, making processes more environmentally sustainable. To achieve this, there are a multitude of strategies, which could include simple energy savings, reduction of greenhouse gas emissions, or simply carefully selecting materials and preventing them from going to landfill through reuse.

Over the course of 10 chapters, behaviors will be described, suggesting policies, and pointing out strategies in the different sectors of activity. From agriculture, which uses countless chemicals with a strong environmental impact, to services, which can save energy and reduce the consumption of some raw materials, such as paper, there is a range of solutions that can help us rethink the way we live, and how we relate to the manufacturers of the products we consume. In addition to policies, which should be radically revised, a strong commitment to environmental education, and a standardization of behavior on the positive side is necessary for each of us to be an active part of the change that needs to be produced in relation to the environment.

The contribution of automation to the environmental cause is fundamental. This can be seen right from the start in our daily lives, at home, where the automation of various devices could lead to considerable energy savings. In construction, it is possible to use automation as a valuable aid to save energy and resources. In industry, the role of automation already represents an undeniable contribution to environmental sustainability, but it can go much further.

It is hoped that the reading and discussion of this book will produce environmentally friendly results, as it is definitely necessary.

The author would like to leave here a very special acknowledgment to his wife, Emília Silva, for all the time and attention that was not dedicated to her during the writing of this book, and also for her eternal affection. The author would also like to leave here an equally special acknowledgment to his daughter Sofia Silva, and her boyfriend, Pedro Nascimento, for their fabulous contribution in reviewing the texts.

Contents

Preface iii

1. Green Principles 1
2. Sustainable Development Goals and Correspondent Policies: A Brief Discussion 17
3. How Can Automation Make the World Greener 36
4. Automation Applied to Manufacturing Systems 74
5. Smart Manufacturing as a Way for Greener Processes 95
6. Additive Manufacturing as a Greener Way to Production Systems 116
7. Advanced Coatings as a Greener Production Process 140
8. Greener Technologies in Other Manufacturing Processes 163
9. Green Automation in our Family Life 182
10. Case Studies 202

Index 221

Chapter 1

Green Principles

1.1 Population and consumerism growth

Given that humanity faces problems that are extremely difficult to overcome, it is important to highlight the biggest sustainable development challenges that several developing countries face: population growth, public health, vulnerability to climate change, human development, and economic growth. However, these problems are not observed exclusively in developing countries. In fact, in 2022, the world's population exceeded 8 billion people. Fifty years ago, the world's population growth rate was around 2%, and currently is around 1.0% [1], which is good news regarding environmental sustainability. However, the population keeps growing. Population growth, by itself, already represents a challenge for global sustainability, as more goods need to be grown and produced to feed these people, which consumes territorial space, natural resources, drinking water and energy. In fact, the planet is no longer able to regenerate everything we consume. Considered on an annual basis, in 1971 the population depleted the resources naturally regenerated by the planet very close to the end of the year, i.e., the planet could regenerate almost everything that the population consumed, in terms of natural resources [2]. However, with the increase in population and consumption, the latter reflecting an increase in the population's average income, the situation has progressively deteriorated. In fact, in 2000, the world's population had exhausted in less than 9 mon the resources that the planet can regenerate in a year. In 2022, the resources that the planet manages to regenerate in a year were consumed in less than 7 mon. This progression is, to say the least, alarming.

However, population growth brings other challenges, through consumption and waste generation, as well as greenhouse gas emissions. Over a 10-yr period, CO_2 values increased from 388.62 ppm in 2010 to 413.27 ppm in 2020. Since the beginning of the industrial age in 1750, human activities have increased CO_2 concentrations by 50%. This increase is more than what occurred naturally over a period of 20.000 yr [3]. The continuous increase in GHG (GreenHouse Gas) emissions has been one of the most serious problems for sustainable development in the 21st century. As vulnerabilities to climate change are shaped by factors such as population dynamism and economic status, as well as the ability to adapt to these changes. People in less developed and developing countries are more likely to be more vulnerable compared to those in developed countries. It is well-known that the United Nations defines sustainable development as "meeting the needs of the present without compromising the ability of future generations to meet their own needs". Sustainable development presents two fundamental educational challenges. One is to promote informed decisions that lead to sustainability. The other is to teach the benefits of integrating the conservation of the planet with the need for economic development. Thus, education must be an extremely relevant indicator in the fight against climate change, and in the reduction of barriers that prevent the achievement of sustainability goals. All trends in terms of the environment and global sustainability show a strong negative trend. To revert this situation, it would be necessary for the countries to reach an understanding to drastically reduce all the causes that are at the beginning of the deterioration of climatic conditions. Reflecting on what is happening today is atrocious, but if we consider the trends, it will be very easy to conclude that the planet that we will leave to the new generations is heavily mortgaged.

With climate change becoming increasingly difficult to ignore, the word sustainability has become fashionable. In fact, sustainability has become an incessant part of the lexicon of politicians, who find it difficult to make unpopular decisions, as they depend on the vote of their constituents. However, marketing knows how to make the best use of the word to try to sell more and at a better price, not always doing its best to earn consumers' trust. Purchasing sustainable jeans or an organic cotton sweatshirt, or any other good with the 'bio' prefix, has become a great way for many brands to capture the preference of their consumers who are more concerned about the environment, and even the most unwary who, without knowing exactly what is behind the words used by marketing, feel that they are making some contribution to saving the environment.

It is a fact that there are brands of the most diverse products making significant efforts to minimize the impact of their activities and their products on the environment. However, legislation should be clear and standardize information so that consumers can be fully informed about the environmental impact ~~that~~ of the product they are purchasing really represents. The life cycle is a terminology used primarily by people with higher education degrees, or with an above average general culture. It will be difficult to transmit assertive messages to those who do not master the basic concepts, hence it is very necessary to invest in education in general, specifically focused on the environment. Product labeling should obligatorily mention the ecological footprint of each product, and governments should promote adequate monitoring of companies for producing reliable information. However, this works essentially in societies where the standard of living is minimally acceptable. On the other hand, how is it possible to demand from populations that struggle to have the essentials, that they need to be critical about what they are purchasing?

The standards of many populations, mainly in countries with medium or high incomes, lead populations to consume, as the basis of economic growth, feeding a chain that generates employment, income, and more consumption. However, many companies look for certain opportunities to innovate in certain products, resorting intensively to industrialization. As an example, milk could be studied, since it is a natural product, but presented to the market in several ways. From milk, a series of other products were created with different presentations, energy value and taste, manipulating the natural product and adding value to it. Once again, marketing does its job, inducing needs even where they apparently should not exist and, with a new presentation, milk can be sold in countless ways, from yogurt to cheese, passing through other much more complex products. The industry does its part: it adds value and offers something that the customer comes to prefer, regardless of the environmental impact that the entire product processing process may have on the environment.

Due to a general increase in income that has been confirmed, the population tends to consume more, mainly what is harmful to the environment: excessive consumption of water, significant consumption of energy, the production of effluents, massive use of packaging and eventual hasty disposal of some products by consumers. It is perfectly legitimate that human beings tend to feel comfortable with themselves when they have the economic conditions to do so, but excessive and uncritical consumption is extremely harmful to the environment. Explicit and implicit advertising is another factor that usually acts negatively towards sustainable practices, as sustainability is still not considered a strong selling point due to the lack of environmental culture of most of the population. Technological evolution also creates pressure on the consumer to change their devices more frequently, with a view to a possible improvement in their operability conditions. Every day we are confronted with novelties in the automotive market, now appealing to sustainability, through electric power-based drive systems. Fashion exists precisely to induce people to change their clothing periodically and rhythmically. And the strategy is repeated in almost all sectors of the market, incessantly appealing to the populations to consume, regardless of the ecological footprint that this may bring. The marketing machine works much more efficiently than any environmental organization warning about the consequences of excessive consumption. It is an uneven fight.

If we consider the joint effect of the sustained increase in population, the increase in their incomes, the pressure of marketing inducing consumption, and excessive consumption itself, all the conditions are met to make humanity walk with great strides towards an unsustainable future.

But can we do anything to avoid this? Yes, it is up to each of us to do our part to reverse the current trend. Everyone's will and the help of technology through automation can contribute in many ways to saving resources and making products more sustainable, greener.

1.2 People, politics, technology, and environment

Prolonged droughts, floods, storms, hurricanes, typhoons, and other natural phenomena have almost always existed, but not with the intensity and frequency with which they have been recorded in recent decades. The Earth's warming record is a very assertive indicator that something is not in the right way. The melting ice at the poles is another undisguised indicator. Humanity has evolved strongly in technological and economic terms, but Man has not been able to properly balance his actions with the need to regenerate nature itself. Our planet is in serious danger, and it is time to put technology at the service of humanity and the planet, minimizing the harm caused and trying to reverse the damage already done. The future is in everyone's hands, but only concerted awareness-raising policies can produce the desired effects.

The world population has been growing consistently, albeit asymmetrically, across different continents. Aggressions to the environment are also carried out asymmetrically, being predominant in the highest population concentrations (urban aggregates), and with greater incidence in the most industrialized countries. Industrialization has two direct implications: most industrial systems consume resources and generate harmful emissions into the environment, on the one hand, it generates wealth that is transformed into the well-being of populations, but also into greater consumption, which corresponds to an equally negative contribution to the environment. In fact, Man yearns for high standards of living, where well-being, comfort, and even some exuberances are essential conditions for him to feel fulfilled. These factors are directly related to consumption. If this consumption was sustainable in environmental terms, it would be perfectly legitimate for humanity to want to enjoy an extremely comfortable standard of living, this standard being different from country to country, and even from continent to continent. However, we know that we consume much more than nature can regenerate, often after about 5 mon in a year, we have spent all that nature is capable of regenerating for everyone. Nevertheless, the selfishness of the human beings prevents, in most cases, that the individual is willing to give up his well-being to the detriment of a more balanced and sustainable environment. Moreover, the prosperity of countries is almost invariably measured by their Gross Domestic Product (GDP), which is based on the generation of wealth, which is largely dependent on consumption. In political terms, it is frustrating for any government not to present a decent growth in a gross domestic product, as it gives the impression that it is not improving the living conditions of its population. This way of thinking has invariably been superimposed on all concepts of environmental preservation. The destinies of the planet continue to be ruled by the economy regardless how loud the environmentalists manifest themselves. The personal, business, and political selfishness is overlapping all the environmentalists' cries of revolt. Seducing citizens with a better economic life may be suggestive of having a much greater capacity of persuasion than defining and complying with restrictive policies of damage to the environment. It is easy to think that, in the short term, we will benefit much more from an economic improvement than to exert some sacrifice for a better future.

World population growth and consumerism are the two main reasons for concern regarding the environment. Feeding a larger population implies taking more advantage of natural resources or using artificial methods to increase food production [4]. These artificial methods of food production cause damage to the environment. Despite the differences between the supply and demand for food on the market can lead to waste, representing the consumption of resources, which must be judiciously spent. The other source of concern is the exacerbated consumerism, which leads to a double waste: the consumption of resources to make the goods available, and the need for regeneration, recycling, or landfill of the goods when the owners understands that they are no longer needed. Both factors could be avoided through adequate environmental education, based on principles different from those that guide distinct types of society. Will it be easy to educate the population of certain countries to procreate in a more moderate way, given the lack of resources in those countries? Is it possible to reduce the number of consumerists in societies that base their status on appearance and on the complete waste of resources? There exists clearly a lack of balance and common sense in the way society faces the future. The present is clearly lived without worries about the future. In some societies, regardless of the conditions that may exist in the future to survive, procreation is something that is in their genesis. For other societies, luxury and waste are the only way of they feel comfortable and good with life. Will it be easy to change this mindset? Politics and certain sectors of society could play a decisive role in the education and regulation of these factors that lead to huge asymmetries, but that also lead to a degradation of the environment that surrounds us, being decisive for our future, and even more for the future of subsequent generations.

The indifference of the population to climate change and the above-mentioned factors are not widespread, being more evident in the more educated population groups and countries where a greater concern for the preservation of the environment is cultivated. However, the melting of the glaciers, the localized and more aggressive atmospheric phenomena, and even the constant warnings from scientists about the problems that we are already facing in the present, but that will certainly worsen in the future, have not awakened the due attention in the general population. Thus, the small acts of environmental aggression and carelessness with the waste of resources are still a constant. The contempt of many populations for excessive consumption of water in situations that would be perfectly avoidable and where no reuse or purification process is carried out, as well as the disdain for energy consumption, which can be easily seen by many offices in skyscrapers where no one is working at night, but that remain with all lighting on, are clear signs that the awareness of the population in general is far from being achieved, and the necessary effort is not being made to reverse this situation. However, political decisions could take a decisive role to reverse this situation, but politicians' fear of pinching the status-quo and ~~the~~ habits in general of certain groups of the population, means that there is not the necessary courage to face current problems, even knowing that they will be much more serious in the future.

In fact, if citizens have the power to control themselves in the different aspects that can contribute to a non-sustainable future, when they do not, they must be regulated by political rules, which have the power to guide behavior in the intended direction, thought as more correct. Considering that each decision in a certain direction always promotes contestations and attempts to circumvent the orders emanating, no one like the political power is able to alert to certain negative factors and force behaviors to align in the sense that the attitudes of its citizens are in accordance with sustainable practices and principles, as perfectly defined by the UN [5]. Certain countries have taken the decision to make the water supplied to homes and businesses more expensive, inducing to reduce its consumption. This is a coercive measure that had an immediate impact on consumption, serving as a more vivid alert about the need to pay attention to the consumption of that good.

In addition to global environmental education, more restrictive policy decisions are needed to prevent a constant aggression to the environment. Current legislation is significantly restricting the use of plastics, with special preponderance in Europe so far. The tendency to abandon plastic packaging anywhere, especially in water courses, is well known. The difficulty in separating and recycling all types of plastic is also known. Furthermore, its use after recycling is limited to certain applications and a certain number of cycles, with the degradation of its properties. The effort to exchange plastic packaging for paper packaging, which is more easily recyclable, has produced very significant results, although there are sectors of less developed societies that insist on not wanting to understand the absolute advantage and necessity of this exchange. If difficulties are felt at this level, even greater difficulties will be felt in more drastic changes that urgently need to be made. Therefore, only a well-designed and explained policy for the populations can effectively reduce the aggression to the environment.

An example of the political conduction of processes for the benefit of the environment, but which nevertheless raises immense doubts, is the change in the energy of vehicles. Nowadays, an immense political effort is being made by several countries to induce the transition from combustion engines to electric powered engines in the automotive industry. However, people question the effectiveness of these measures in environmental terms. In fact, the way electrical energy is generated, and storage are factors that still raises enormous controversy. If the energy is generated based only on renewable sources, half of the problem is solved. But, will we have the capacity to generate all the electrical energy needed for the mobility that is expected to exist in the coming decades? In political terms, too, a huge effort needs to be made to make this possible. Additionally, batteries pose another problem. The life cycle of the materials used in batteries, as well as the batteries in themselves, will have to be properly studied and considered. The reduction of atmospheric emissions does not translate directly into the enslavement of peoples, the pollution of certain areas where these materials are abundant. It is necessary that the end of life of these batteries does not become a nightmare for humanity. However, when citizens do not express an interest in adapting to new environmental needs, it is up to governments to adopt measures that show their citizens the way, both through the obligation to follow certain rules and through interference in market prices that lead to citizens tending to adopt more economically rational choices, and that these are in accordance with the general guidelines of environmental sustainability that a given community should follow, with a view to the common good of humanity.

What is currently happening with the change in vehicle locomotion is also closely linked to technological factors. Technology has acted in both positive and negative ways towards the environment. In the first phase, industrialization relied heavily on technology to develop production systems, and this development was carried out at any price. In fact, economic growth allowed for greater purchasing power, which translated into a greater need to supply all kinds of products to the market, mainly based on the industrial side. The industry is largely based on technology, which, in this initial phase, was focused only on increasing production, without environmental concerns. Only after science alerted to the harm that was being done to the planet, that humanity woke up to the need to put technology at the service of the environment. In the past decades, technology has been a strong ally of humanity in the sense of allowing the ecological damages to be minimized or even eliminated. The term Cleaner Production has become much more common, and government restrictions have increased significantly regarding pollution. However, human beings are quite creative and have adopted some practices to overcome these rules, transferring the most harmful processes to countries with large employment needs and relatively cheap labor, only transferring the harm produced to the environment from place to place. Countries that need to install multinationals to create jobs for their population usually have a low bargaining power, allowing investors to save on the initial investment in technologically advanced systems that minimize negative effects on the environment. In countries where environmental awareness is more ingrained, development is essentially carried out around technology capable of mitigating the possible negative effects on the environment of each type of industry. The developments achieved in this area are remarkable, but they can only be accessed by-the more economically favored countries, and where environmental concerns are more rooted. Thus, "greener" has assumed special attention in some countries and cultures, but this costs money and the consumer needs to be aware that the use of technology overcharges prices, but are the only way to get a better future.

1.3 Green concept

1.3.1 Green as part of sustainable concept

Green and sustainability are concepts that are usually mixed, the first being clearly embedded in the second. Over time, sustainability has gained a much broader spectrum, including three aspects [6]: economic, environmental, and social. Economic sustainability has become a vital concept for companies to be viable and is closely linked to their competitiveness in the market, which is also associated with factors such as innovation capacity, product quality, target markets, delivery capacity, timely delivery of products according to the commitments assumed, ability to capture talents and specialized workforce, processes conveniently define, among many other factors. The awakening of humanity to environmental issues and the fact that some companies wanted to achieve their economic sustainability at any price, created the development of the concept of environmental sustainability, which evolved immensely in the last decades of the 20th century, triggered an immense standardization and regulation. The social issue also began to become important, considering all the social factors that revolve around the need to subsist, but taking into account human dignity, avoiding unnecessary risks, as well as occupational diseases that come from contact with substances or dangerous environments, extremely repetitive tasks or in very unergonomic positions, which lead to unnecessary fatigue.

The concept of 'green' comes essentially from the fact that human beings feel good about themselves and nature. Any conscious human being is aware of the danger that certain environments or processes can bring to himself and to humanity. A farmer feels the need to protect his crops against certain diseases that can affect his crop, but he can do it in different ways: using products that, in addition to having the intended effect, are going to rivers, lakes, or the sea and polluting these waters, or using natural products less harmful to the environment. This is just a case concerning the primary sector, but it can easily be replicated in other sectors of economic activity. Only a greater knowledge of the environmental impact of products and a greater awareness of preserving nature can lead to 'green' attitudes, of replacing environmentally aggressive products and processes with others that induce a lower environmental impact. After much harm induced by humanity to the planet, we live in the industrialization phase, and only after detecting the effect of these same harms on the environment, with the consequences that we are all feeling, studies were started that allowed the identification and quantification of these effects, allowing the study of their causes and their corresponding mitigation. This mitigation, and its corresponding practices, ended up leading to what is conventionally adopted as 'green' practices, which lead to reaching the same objective, or very close, using other products or processes, which have less impact on the environment.

1.3.2 The Green concept applied to some important sectors

To emphasize certain ideas, the scientific community has the practice of labeling certain aspects of development, to demarcate these studies from the reactants and allowing better identification and differentiation of the same. Green was the designation chosen for development studies aimed at concepts that allow for a better environment. The term arose from the concept of comfort, well-being, and serenity that humanity normally feels in contact with nature, and the color of lawns and meadows is particularly prominent in our subconscious. Projects and companies also began to preferentially use the color green in almost everything that is closely linked to improving environmental sustainability. Since then, particularly the scientific community, has designed numerous models that aim to cement the green concept in various aspects of knowledge, from the primary sector, with the use of fertilizers and pesticides that are less harmful to the environment, to the industry, which tries to reduce as much as possible the consumption of natural resources and starts to develop products duly studied in terms of life cycle.

The construction sector has been pointed out as a sector with outstanding scientific activity. In fact, buildings are infrastructures that are intended to be durable, and in which any design error in environmental terms is not always easy to resolve *a posteriori*. And if it does, it usually brings considerable costs. Poor thermal insulation leads to excessive energy consumption, especially in the hottest and coldest seasons, and this consumption will continue until improvement works are carried out, or the building is demolished. Construction started to take more into account the sun exposure, taking advantage of it, and avoiding insulation gaps where this exposure was not so favorable. In terms of lighting, although energy consumption is not normally as high as in the thermal component, it can create discomfort and the need for artificial lighting. Thermal and lighting issues can contribute to the environment in a very harmful way or allow a sustainable concept to be followed. Hence, studies focus a lot on this area. The use of materials, if judicious and taking into account the environment, can also minimize processing and avoid the consumption of certain natural resources. The use of materials that are more difficult to degrade will allow longer use without the need for reconditioning, saving resources. The use of materials that considering their function, if well selected, can help in a more sustainable way to achieve the intended goals. Materials with simpler processing, which do not involve the use of harmful products for the environment and do not excessively consume natural resources (water, energy, etc.), can be a perfectly viable solution, to the detriment of others that consume more resources and that may be difficult to recover, reuse or recycle at the end of their useful life cycle. In fact, as with any product, buildings should be designed considering their life cycle, trying to minimize all possible harmful effects that they may have on the environment. These are some vectors explored by several researchers and that have led to quite significant results in terms of increasing sustainability in this sector, with benefits for the whole of society.

Another sector where the "green" concept has been applied for some decades is the chemical industry. This industry covers many sectors, from oil to food, as well as numerous reagents that are used in other products as by-products. This industry is also responsible for the manufacture of pesticides and fertilizers for agriculture, but also reagents, solvents, paints, varnishes, and several other products known to be harmful to the environment. In addition to the danger to the environment and public health associated with the processing of these products, their life cycle has become the subject of numerous studies, given the threat to the environment and humanity they represent. In fact, the chemical industry was responsible for several accidents over many decades that brought significant damage to the environment, due to explosions, leaks or spills of harmful products, serious occupational diseases, among others. Thus, the concept of "green" for the chemical industry has a different character from what is normally attributed to the construction sector. In fact, in industry, it has become important to drastically reduce the production and consumption of products largely harmful to the environment, drastically reduce effluents or treat them properly, filter certain gaseous emissions, and reduce the generation of gases with the greenhouse effect. Although the needs in this sector are diverse, they are still as or more important than the concerns in the construction sector. The various forms of pollution generated by the chemical industry became an even greater concern in the last decades of the 20th century, due to the increasing processing of the most diverse products of a chemical nature, which potentially bring high risks to the environment when released into nature without proper treatment. As environmental and public health accidents are usually significantly noticeable, environmental awareness in the most developed countries in terms of environmental sustainability, and the legislation imposed in the meantime on this type of industries has generated two phenomena: (1) flight from large centers to minimize possible harmful effects of emissions, waste generation, or serious accidents to public health; (2) installation of this type of industries in countries more permissive in terms of legislation, and where society and environmental organizations are not opposed to any kind of environmental negligence and excessively high risks to public health. The trend of the development of products that use less harmful elements has become a current practice within the most diverse aspects of the chemical industry, which began to find economically viable alternatives, which entailed much lower risks for the environment and health of the public. Announcing these advances has also become a banner for the most prominent companies in the chemical industry, which has always had a strong associated research and development component.

Although the chemical industry has a prominent place in global environmental concerns, in general almost the entire industry contributes to environmental degradation in different aspects and levels: it consumes natural resources, releases effluents that can be relatively harmful to the environment, and produces products that can have different degrees of reuse, recycling, or disposal environmental impact. Regarding the consumption of natural resources, numerous theories have been developed and several studies have been conducted to reduce or eliminate the inclusion of products hazardous to health and the environment, as well as minimizing the consumption of certain types of natural resources that have greater difficulty of regeneration by nature. Furthermore, and of crucial importance at the time, in the years 2022 and 2023 in which this work is being written, the energy consumed is an increasingly crucial asset, given that about two-thirds of the electrical energy currently consumed is still being generated based on fossil fuels [7]. Effectively, nowadays, and in terms of energy, there are two sustainability concerns: environmental sustainability through the reduction of greenhouse gas emissions, and economic sustainability, due to the monopolies of some natural energy resources and the use of these monopolies for political and geostrategic purposes. In this regard, although many governments have already taken extremely positive actions in promoting renewable sources for electricity generation, these steps have not been vigorous enough to save the environment and reduce energy dependence on fossil fuels. Fortunately, there are very positive steps already taken by several countries in encouraging renewable energy generation facilities for self-consumption, namely, through economic support for the installation of solar panels. Favorably, some companies are also adopting this strategy to reduce costs and contribute to a greener environment. However, in addition to government policies, which signal and point out the necessary way to follow, there must be an awareness of the population so that, in addition to industry, each citizen also contributes to this goal. Electric power generation is one of the industries that contributes the most to environmental degradation and, since there are already perfectly solid alternative paths that allow achieving the same energy production without degrading the environment, it is urgent to develop assertive policies for generating energy essentially through renewable means. Effectively, in terms of energy, the world must do more and better, and this involves mitigating the causes of excessive energy consumption, trying to produce as much energy as possible for self-consumption, and being judicious in the energy sources used by the different energy suppliers. These will be the main aspects to make the activities of each company greener.

Products manufactured by the industry are primarily developed based on market needs and expectations, to outperform other existing products in terms of performance, filling duly identified gaps, or creating new consumer needs. On the one hand, the industry tries to find ways to attract the market, processing products in different ways, which could be consumed purely and simply in their original form. An example of this is all the ways in which fruit can be presented to consumers, from the dehydrated form to inclusion in yogurts, among many other ways. On the other hand, there is the consumer who, faced with a new way of presenting a given product, sees it as an advantage for reasons of personal taste, ease of acquisition, comfort in use, etc. However, the industry adds value to a given initial product, which is favorable in terms of economic and social sustainability (generates employment) but consumes more energy and other by-products that can even be harmful to the environment.

This is a duality of interests that could easily have a smaller effect, that is, being greener, if consumption were to decrease, a situation that goes against the economic development that politicians want so much to compensate for the increase in expenses they create. Product development has two curious facets, among a few others: it is compelled by marketing to increase consumption and attract market attention, and it can be a way to make the industry greener. Marketing is not exactly the aspect of this work, thus, attention to how the product can become greener during its development phase is paid. Indeed, there are numerous studies on new product development guidelines and strategies [8-18], which is also extended to product distribution logistics [19-20]. Depending on the strategy adopted by each company, product development could follow relatively strict lines in some aspects, such as the preferential use of natural products, the non-use of potentially hazardous products for health and the environment, how the product will be produced, the durability of the product and how it can be reused, recycled, or discarded. However, some products can even cause damage to the environment even during their use, or during their useful life, a factor that must also be considered in the product development phase. In fact, many studies have been focused on accessing the life cycle of products, quantifying their impact in environmental terms [21-30]. If the resources spent in the manufacture of products and the effluents caused by the manufacturing processes are a serious problem for the environment that has been mitigated over time, either by the companies' own efforts or by legal impositions, the end of life of the products also represents a serious problem for the environment in most cases. Plastics, despite a part being recyclable, have caused numerous problems worldwide, and it is now being observed that the presence of microplastics has reached the human food chain. In fact, plastics can break down into particles small enough to pass through water filters. Given that microplastics easily reach water courses, then they reach the oceans, where marine species absorb them, culminating in the insertion into the human food chain. This is just one of the many problems caused by countless products, made up of the most diverse materials that, when abandoned at the end of their useful life, continue to constitute a serious problem for the environment. The ways to mitigate this problem are to contribute to the decrease in the use of these materials in the development of new products, and to increase the collection of products at their end of life, control their environmental impact after they are no longer useful. Recycling is an extremely interesting way, not always economically viable, but which should be explored whenever possible, avoiding the landfill of more products harmful to the environment. However, the first step should be taken precisely in the development of the product, not allowing the user to dispose of the product with damages to the environment previously known by those who manufacture the products. To solve these problems, it is necessary for a large increase of the environmental conscience in the training programs of the future professionals linked to the development of products, education of the top management so that they do not let themselves be influenced by greed, applying materials harmful to the environment just because are more economical or easier to transform, and that consumers have a greater environmental culture that allows them to consume less and know how to select products that have a lower ecological footprint during their manufacture, and which will lead to a lower environmental impact during their production, its use and disposal. Education and training are some of the keywords to become our world greener.

In industries, another major problem that needs to be mitigated is the generation of effluents. A part of the transformation processes involves the processing of raw materials, some of which are harmful to the environment as by-products. The processing of some products involves high temperature processes, conditions that usually lead to the release of vapors that can vary from perfectly harmless to extremely dangerous. Furthermore, numerous manufacturing processes use significant amounts of water, which causes two problems: consumption of potable water, on the one hand, and the release of contaminating water, on the other. Both in terms of solid, liquid, and gaseous effluents, the legislation has imposed severe rules that, if not complied with, normally imply the application of severe fines, which can make the operation of certain non-compliant companies unfeasible. However, as already mentioned, not all countries have legislations capable of dissuading companies from violating the environmental aspect. Thus, there has been a migration of certain types of industries from developed countries to other developing countries, where legislation is more permissive or there is not such effective control. This situation only transfers the problem of location, but within the same planet, the one on which we all live. This problem also has two ways of solving it, which have already been followed by many companies: designing products based on different raw materials and processing processes, which do not harm the environment, and harmonizing environmental legislation in world terms, through a greater commitment by developing countries, so that they do not accept to host companies that are known by not respecting-the environmental rules already established in developed countries.

1.4 Environmental education for a greener world

Despite the enormous effort made by environmental organizations and the political rhetoric of many governments, the environment remains a marginal issue for many, and even completely ignored by others. The pace at which the harm imposed on the environment has led the world to a situation close to no return. The Secretary-General of the United Nations has made incessant appeals for adopting policies capable of reversing global warming with the intention of drastically reducing greenhouse gas emissions, decreasing the generation of electricity based on highly polluting fuels such as coal, and the less use of disposable packaging and bags, to avoid massive cuts in forests, among many other conditions in which there is an urgent need for everyone to intervene, from governments to common citizens. These calls, already almost in a desperate tone, were heard again during the COP27, UN Climate Change Conference, in November 2022. This conference took place in an atypical period, occurring after a pandemic period in which the reduction of emissions was a fact, as a result of the confinement of people, as well as the closure of part of the industrial activity and tourism. However, in 2022, pollution rates are practically at the same level as in 2019, that is, this period was not used to learn, reflect, and change. The conflict situation experienced in Europe between Russia and Ukraine led to an energy crisis, which led certain European countries to regress some electricity generation policies. The use of highly polluting materials in the production of electricity has begun to be considered to compensate for the shortfall in gas and oil supplies from Russia. If there is evolution in a positive direction, it is so tenuous that it goes unnoticed. It gives the feeling that there is always something more urgent to deal with than improving the environment. And so, we consistently walk towards chaos.

In addition to environmental groups, some groups of young students are beginning to realize that the world in which they will have to live is striding towards this chaos. The rise of sea water, aggressive weather phenomena, drought and strong fires will become increasingly common. The perception of these facts becomes frightening for this young population, which, even in a somewhat insipid way, begins to wake up to this reality. Activist Greta Thunberg has been the face of this dissatisfaction-and is beginning to reign in the younger strata. But where are the teachers of these young people, and in what way are they helping this young layer to understand the chaotic situation we are heading towards?

Environmental awareness must start within each family, through daily and consolidated practices that aim at the global increase of sustainability, not only economic, but essentially environmental and social education. However, new generations will be educated by other generations who contributed to the current state of the environment. Thus, deep reflection will be necessary. However it is difficult to believe that the new generations educate their parents, something needs to be done so that awareness spreads as quickly as possible. In fact, it is the young generations that see their future most threatened and have raised their voices to influence politicians and decision-makers to take actions. Therefore, the initiative is in counter-cycle with what we understand by education, that is, the youngest are asking adults to respect rules they consider basic so that the environment does not continue to deteriorate at the rate we have been witnessing.

Education is a key factor that can promote the necessary change. Probably, the new generations will have to speak out loudly for decision-makers to establish new rules that prevent environmental degradation. On the one hand, decision-makers place much more attention on economic aspects than on environmental policy. On the other hand, economic power exerts enormous pressure on political power, containing the impetus for reforms to which governments could be susceptible, due to popular protests. Although necessary, change can never be radical. Changing from a combustion engine paradigm to an electric motor paradigm would correspond to obvious and dreadful-problems: (1) the electrical grid would not be prepared for such an overload of electrical energy consumption, both in terms of generation and distribution, which could lead to local or regional blackouts; (2) combustion engine manufacturers need to adapt to the production of new products, under penalty of massive unemployment, since the skills required by one industry do not exactly match with those needed by the other. This is just one example of the care that politicians should take in implementing the necessary changes.

Thus, despite education being a key factor in the development of environmentally friendly policies, the inertia imposed by economic and political power can only be overcome by the market itself. For this, the market must have the necessary environmental awareness. Only awareness and fear of approaching dark days in terms of climate change will be the necessary change factors for the situation of environmental degradation to be reversed. Although the awareness of some older generations is a reality, it is easy to think that the drastic changes that can happen will no longer affect them in particular. This state of thought leads to the current inertia. Only then will the younger generations will be able to reverse the situation, which goes against the normal flow of information and education, but which may be the only one that works successfully in this case. Activist organizations will have to strengthen themselves and promote more actions and greater pressure on governments, so that environmental awareness becomes part of the curricula in all age groups, given the urgency of reversing the situation, but considering the need to sensitize populations from an early age. The media needs to be much more involved in these awareness programs, to refute the economic interests that manipulate governments, preventing more demanding regulations from being imposed on companies regarding environmental aggressions. Unfortunately, regulations are not effective, they are still quite scarce, and they are applied very unevenly across different areas of the planet, leading to asymmetries in competitiveness caused by imposing these same restrictions. Thus, education plays a key role in the creation of products, production processes and consumption habits that contribute in a more positive way to this problem that already affects us significantly, but which can be catastrophic for future generations if behaviors are not reversed in the meantime. Information capable of showing the effects and directly correlating them with the causes is needed, drawing everyone's attention to the importance of all our daily actions and decisions in this process.

References

[1] Roser, M., Ritchie, H., Ortiz-Ospina, E. and Rodés-Guirao, L. (2022). World Population Growth. URL: https://ourworldindata.org/world-population-growth, accessed on November 30th, 2022.

[2] Earth Overshoot Day. (2022). This year, Earth Overshoot Day fell on July 28. URL: https://www.overshootday.org/(Accessed on November 30th, 2022).

[3] NASA. (2022). Carbon Dioxide | Vital Signs – Climate Change: Vital Signs of the Planet. URL: https://climate.nasa.gov/vital-signs/carbon-dioxide/(Accessed on November 30th, 2022).

[4] Silva, F. J. G. and Gouveia, R. M. (2020). Sustainable production cases. pp. 281–373. *In:* Silva, F. J. G. and Gouveia, R. M. (Eds.). Cleaner Production – Toward a Better Future, Springer Nature Switzerland, Cham, Switzerland, 2020. doi: 10.1007/978-3-030-23165-1_8.

[5] UN. (2022). UN Sustainable Development Goals, retrieved from: https://drc.ngo/about-us/who-we-are/un-sdgs/?campaign={DRIP-English}&adgroup={All-pages}&creative=490088684572source={google_ads}&medium={Grants_google}&keyword=&content={content}&time={2022}&gclid=Cj0KCQjwkt6aBhDKARIsAAyeLJ1Xy-jpuATATlCoTIt8Y1ivhhU81ZBCI9_iMTaZwkqqnHYGf2vPSNIaAm_pEALw_wcB, (Accessed on October 25th, 2022).

[6] Silva, F. J. G., Kirytopoulos, K., Ferreira, L. P., Sá, J. C., Santos, G. and Nogueira, M. C. C. (2022). The three pillars of sustainability and agile project management: How do they influence each other, Corporate Social Responsibility and Environmental Management, 29: 1495–1512. doi: 10.1002/csr.2287.

[7] Our World in Data (2022). Electricity mix, retrieved from: https://ourworldindata.org/electricity-mix, (Accessed on November 7th, 2022).

[8] Watz, M. and Hallstedt, S. I. (2022). owards sustainable product development – Insights from testing and evaluating a profile model for management of sustainability integration into design requirements, Journal of Cleaner Production, 346: 131000. doi: 10.1016/j.jclepro.2022.131000.

[9] Schulte, J. and Knuts, S. (2022). Sustainability impact and effects analysis - A risk management tool for sustainable product development, Sustainable Production and Consumption, 30: 737–751. doi: 10.1016/j.spc.2022.01.004.

[10] Wan, C. K. and Lin, S.-Y. (2022). Negotiating social value, time perspective, and development space in sustainable product design: A dialectics perspective, Design Studies, 81: 101121. doi: 10.1016/j.destud.2022.101121.

[11] Wang, S., Su, D., Ma, M. and Kuang, W. (2021). Sustainable product development and service approach for application in industrial lighting products. Sustainable Production and Consumption, 27: 1808–1821. doi: 10.1016/j.spc.2021.04.003.

[12] Wang, Y., Modi, S. B. and Schoenherr, T. (2021). Leveraging sustainable design practices through supplier involvement in new product development: The role of the suppliers' environmental management capability. International Journal of Production Economics, 232: 107919. doi: 10.1016/j.ijpe.2020.107919.

[13] de Guimarães, J. C. F., Severo, E. A., Jabbour, C. J. C., Jabbour, A. B. L. S. and Rosa, A. F. P. (2021). The journey towards sustainable product development: why are some manufacturing companies better than others at product innovation? Technovation, 103: 102239. doi: 10.1016/j.technovation.2021.102239.

[14] Santos, J., Gouveia, R. M. and Silva, F. J. G. (2017). Designing a new sustainable approach to the change for lightweight materials in structural components used in truck industry. Journal of Cleaner Production, 164: 115–123. doi: 10.1016/j.jclepro.2017.06.174.

[15] Garcia-Garcia, G., Azanedo, L. and Rahimifard, S. (2020). Embedding sustainability analysis in new food product development. Trends in Food Science & Technology, 108: 236–244. doi: 10.1016/j.tifs.2020.12.018.

[16] Hafezi, M., Zhao, X. and Zolfagharinia, H. (2023). Together we stand? Co-opetition for the development of green products. European Journal of Operational Research, 306(3): 1417–1438. doi: 10.1016/j.ejor.2022.07.027.

[17] Kono, J., Ostermeyer, Y. and Wallbaum, H. (2018). Investigation of regional conditions and sustainability indicators for sustainable product development of building materials. Journal of Cleaner Production, 196: 1356–1364. doi: 10.1016/j.jclepro.2018.06.057.

[18] Held, M., Weidmann, D., Kammerl, D., Hollauer, C., Mörtl, M., Omer, M. and Lindemann, U. (2018). Current challenges for sustainable product development in the German automotive sector: A survey based status assessment. Journal of Cleaner Production, 195: 869–889. doi: 10.1016/j.jclepro.2018.05.118.

[19] Fung, Y.-N., Chan, H.-L., Choi, T.-M. and Liu, R. (2021). Sustainable product development processes in fashion: Supply chains structures and classifications. International Journal of Production Economics, 231: 107911. doi: 10.1016/j.ijpe.2020.107911.

[20] Lopes, P. V., Silva, F. J. G., Campilho, R. D. S. G. and de Almeida, F. (2019). Designing a novel and greener truck asphalt container. Procedia Manufacturing, 38: 324–332. doi: 10.1016/j.promfg.2020.01.042.

[21] Subramanian, V., Peijnenburg, W. J. G. M., Vijver, M. G., Carlos F. Blanco, Cucurachi, S. and Guinée, J. B. (2022). pproaches to implement safe by design in early product design through combining risk assessment and Life Cycle Assessment, Chemosphere, 311(1): 137080. doi: 10.1016/j.chemosphere.2022.137080.

[22] Mir, N., Khan, S. A., Kul, A., Sahin, O., Sahmaran, M. and Koc, M. (2022). Life cycle assessment of construction and demolition waste-based geopolymers suited for use in 3-dimensional additive manufacturing. Cleaner Energy and Technology, 10: 100553. doi: 10.1016/j.clet.2022.100553.

[23] Ubando, A. T., Anderson, E., Ng, S., Chen, W.-H., Culaba, A. B. and Kwon, E. E. (2022). Life cycle assessment of microalgal biorefinery: A state-of-the-art review. Bioresearch Technology, 360: 127615. doi: 10.1016/j.biortech.2022.127615.

[24] Dieterle, M., Fischer, P., Pons, M.-N., Blume, N., Minke, C. and Bischi, A. (2022). Life cycle assessment (LCA) for flow batteries: A review of methodological decisions. Sustainable Energy Technologies and Assessments, 53(Part A): 102457. doi: 10.1016/j.seta.2022.102457.

[25] Partanen, J., Kylänpää, M., Loukusa, S., Korkiakoski, M. and Salonen, J. (2022). Device life cycle management requirements for identity and access management in the factory of future environment. Procedia Computer Science, 204: 235–244. doi: 10.1016/j.procs.2022.08.028.

[26] Mio, A. and Fermeglia, M. (2022). Multiscale modelling techniques in Life Cycle Assessment: application to product design. Procedia CIRP, 105: 688–693. doi: 10.1016/j.procir.2022.02.115.

[27] Aguiar, J. B., Martins, A. M., Almeida, C., Ribeiro, H. M. and Marto, J. (2022). Water sustainability: A waterless life cycle for cosmetic products. Sustainable Production and Consumption, 32: 35–51. doi: 10.1016/j.spc.2022.04.008.
[28] Nunes, I. C., Kohlbeck, E., Beuren, F. H., Fagundes, A. B. and Pereira, D. (2021). Life cycle analysis of electronic products for a product-service system. Journal of Cleaner Production, 314: 127926. doi: 10.1016/j.jclepro.2021.127926.
[29] De Wolf, C., Cordella, M., Dodd, N., Byers, B. and Donatello, S. (2023). Whole life cycle environmental impact assessment of buildings: Developing software tool and database support for the EU framework Level(s), Resources, Conservation and Recycling, 188: 106642. doi: 10.1016/j.resconrec.2022.106642.
[30] World Economic Forum, "Microplastics in food chain: how harmful they are?". Retried from: https://www.weforum.org/agenda/2022/06/how-microplastics-get-into-the-food-chain/?DAG=3&gclid=CjwKCAiA9qKbBhAzEiwAS4yeDdhvoC3k3JYTaroQqUGTwbygBXfqI7Gb27Ce6gH-JTPNNqWqxywm1hoCwKAQAvD_BwE, (Accessed on November 7th, 2022).

CHAPTER 2

Sustainable Development Goals and Correspondent Policies

A Brief Discussion

2.1 Is a sustainable development possible?

It is well known by all that the evolution of humanity and the planet as we know it is not evolving the way most of humanity would like. The economic and social imbalances between and within countries is overwhelming, if we compare some countries in Africa, South America, Asia, and even Europe, with others such as the United States of America, Japan, Australia, and most European countries. The different regions of our planet do not have the same natural resources, which could immediately generate considerable asymmetries in the development of their economies and society. However, this factor usually turns out to be clearly overtaken by political factors that come from a geostrategy that is gradually determined by economic and group interests, which easily convert areas with abundant natural resources into regions where communities live in a degraded way. A good portion of the countries with more abundant natural resources are subject to political systems dominated by economic interests, where the wealth generated by these resources is shared by a very restricted group of people, who enrich themselves immeasurably, while miserable living conditions are provided to the remaining population. Although one considers that humanity has evolved immensely, mainly due to technology, it should be forgotten that there are still people without any educational level, and that they only survive due to food aid and health programs. Basically, the so-called more civilized peoples, in addition to wasting money on the purchase of products that are often useless, are helping to ensure that another part of the world's population does not live in such a miserable way, or even perish due to hunger and lack of medical care.

Aware of these imbalances that probably date back to the origins of humanity, the United Nations is trying to act neutral in this attempt at balance, but where this balance is far from existing. In fact, the UN knows that 2 billion people in the world do not have regular access to safe, nutritious and sufficient food. Moreover, about 144 million children under the age of 5 in 2019 were stunted [1]. In 2030, the UN predicts that around 840 million people will suffer from hunger, with special incidence in the African and Asian continents. One of the main causes of the existence of hungry populations lies in the lack of resources to acquire food, which stems from several problems, such as (1) lack of employment, (2) problems of inequality of opportunities, (3) exploitation by networks of organized crime dedicated to modern-day slavery, (4) serious problems of basic education and ability to work, (5) living in areas with poor access to food, (6) living in underdeveloped areas with problems of the generation of employment, (7) residence in places usually affected by extreme weather phenomena, (8) problems of social inclusion, or even (9) inability to accompany the evolution of society in the manner in which it has been stated , among others many other possible causes. Hunger is not only felt in Africa and Asia, it is a "disease" common to almost all countries and populations, not always due to lack of access to opportunities, but also due to bewilderment regarding the way of life that is established in societies. It is common to observe situations of strong food shortages in extremely populous urban areas, where, due to the most diverse situations of a social nature, the search for and achievement of opportunities, inability to work, inadequacy of skills in relation to market needs, and even mental health, among many others, there is the isolation of people who are uprooted from the family and society, self-excluding themselves and falling into situations of serious social precariousness, which reaches the contours of food scarcity. The concept of hunger also encompasses serious food shortages in terms of quality, not just quantity [2]. In fact, human beings need a diversified and balanced diet in terms of nutrients so that their health can be preserved. Due to the many factors mentioned above, not all populations have access to goods considered essential, which clearly contributes to the deterioration of their nutritional balance and health status, also affecting their quality and life expectancy. The UN has played a fundamental role in bringing essential goods to people s traditionally with difficulties in accessing minimum survival conditions by their own means, promoting systematic food aid programs.

UN forecasts for 2023 also indicate that around 6% of the world's population will live in conditions of extreme poverty, estimates that may however be compromised, due to the outbreak of the COVID-19 pandemic experienced in the meantime [3]. Moreover, having a job also does not guarantee that the population has a dignified life. In fact, 7.1% of the world's population that has a job, despite having a job in 2019, also lived in conditions of extreme poverty [3]. Some causes pointed out by the UN for the high degree of extreme poverty existing worldwide are the lack of employment, social exclusion, vulnerability to diseases and atmospheric phenomena or natural disasters, leading these populations to very low productivity rates, which reflected in obtaining the corresponding incomes, generating difficulties in raising goods for their subsistence. However, the documents do not mention other well-known phenomena, such as (1) the excessive ambition of certain sectors of the population, which contrast with the complete lack of ambition of others, (2) the exploitation of some sectors of the population by others, (3) a lack of sensitivity to the economic and environmental conditions in terms of their own sustainability (which translates into lack of social sustainability), (4) the lack of equal opportunities for all, (5) a lack of basic education and exploration of their natural abilities, and (6) the lack of preparation for the technological evolution that is felt worldwide, relegating populations with less access to technological education to a gap in terms of job opportunities and social evolution. These are just some of the many reasons that continue to fuel the problem of huge inequality in worldwide terms.

Additionally, the UN also estimates that in 2030, only about 39 to 63% of the world's population will have effective access to health services [2]. In 2018, about a fifth of the world's school-age population (258 million children and young people) are not attending any type of study. Effectively, in 2018, only about 84% of children completed primary education, which contrasts with 70% that existed in 2000, and it is expected that in 2030, a rate of 89% of children completing primary education worldwide will be reached [4]. In 2018, around 778 million people worldwide, of which two-thirds were women, were completely illiterate, and unable to read or write [4]. Despite some advances made in terms of education, there is still a long way to go, as it is expected that the COVID-19 pandemic during 2020–2021 has delayed the evolution felt, due to the closure of many schools and the online teaching in many countries, which could have negative consequences for the recovery process that was being achieved. The problem of education for women is more serious because about a third of developing countries do not allow equal access to education for children of different genders [4,5]. The fact of being born female can largely condition the course of human life in many countries. Mainly in 30 countries in Africa and the Middle East, about a third of girls aged between 15 and 19 have already been victims of genital mutilation, with all consequences that this can entail for their future life, from a greater probability of contracting infections, or even infertility problems [5]. Gender violence is also a worrying issue, with females predominantly suffering aggression from husbands, boyfriends, or male partners. In fact, around a third of women between the ages of 15 and 49 worldwide have been victims of physical, sexual, or psychological aggression, and this problem does not occur exclusively in poor or developing countries [5]. In many countries in Asia, Africa and the Middle East, there are no equal conditions for access to education for males and females, with greater ease of access for males, and clear restrictions on the female population. This translates into a conditional educational development, which implies fewer opportunities to access employment and a conditioning of their psychic and social development. Effectively, there are still societies where it is established that the male population enjoys numerous advantages over the female population, a situation that creates social imbalance and difficulty for the female population to access many opportunities. This is such a current problem that in 2022, many social struggles are still unfolding to end inequalities between people of different genders in some countries, predominantly in the Middle East.

Drinking water, being one of the essential goods of life, still represents a serious problem due to several aspects. Climate change has created new arid zones, which were previously bathed by water courses. These watercourses had promoted the settlement of certain populations, which, given the new climate change framework, are forced to look for new places where they can supply themselves with drinking water. Water contamination is also a serious problem, mainly due to pollution caused by agricultural and industrial practices inappropriate for the environment. Lack of access to potable water, as well as problems related to sanitation, are at the root of diseases such as malaria and dysentery, which cause millions of victims every year. In fact, 4.2 billion people still face difficulties in accessing basic services such as access to clean water or sanitation. The UN estimates that currently around 673 million people still defecate in the open [6]. Health problems relating to these situations are evident. There is still a clear mismatch between the places where populations are concentrated, the size of these populations, and the infrastructure created to serve these people in the most basic situations.

Energy is almost a basic good as drinking water, given the current state of most societies. In fact, energy is linked to the way we lead our lives, from the mobility we need to the comfort of our homes. In terms of economic activity, which underpins the way society is set up, it is equally fundamental. In 2022, affordable access remains a major issue [7]. Given the still moderate development of renewable electricity generation systems implemented by several countries, fossil fuels continue to be the main energy source. In addition to all the environmental problems that the use of fossil fuels entails, the reserves of these energy resources are located mainly in countries governed by autocracies, with a strong association between this group of producing countries (OPEC – Organization of the Petroleum Exporting Countries), that conditions the market. Despite Ryan Thombs [8] argument that reducing oil dependence may not lead to proportional improvements in the well-being of the population, taking into account a study having as the main target the US population, this dependence can create geopolitical and geostrategic imbalances that could cause serious damage to many millions of people, as it was observed in 2022 and 2023 concerning the situation generated by Europe's dependence on supplies of hydrocarbons from Russia [9]. Energy dependence, armed conflicts, and disasters of various kinds effectively impact energy markets, which can largely affect the comfort of the most vulnerable populations. The transition from the hydrocarbon-based electricity generation systems to renewable energies has been slower than desirable. This could be due to different factors, such as: (1) less environmentally friendly energy production policies; (2) lack of environmental awareness due to a culture based on the production of electricity essentially with fossil resources; (3) difficulty in accessing clean energy production technologies; (4) privileged access to hydrocarbons in a more economical, but less environmentally friendly way of producing energy; (5) lack of infrastructure and skills to deal with new energy production technologies; or (6) lack of economic conditions for the reconversion, even if slow, of energy production systems. Although energy problems affect the entire world in some way, it should be noted that 789 million people do not have access to electricity, 548 million of whom are in Saharan Africa [7]. The pollution generated by the burning of fossil products for energy generation is one of the main sources of pollution. There has been an increased effort in some countries to migrate from systems based on generating electricity using fossil fuels to renewable energy generation systems, using wind turbines, solar panels, harnessing water systems, energy from sea waves, use of biomass, among others. This transition is already

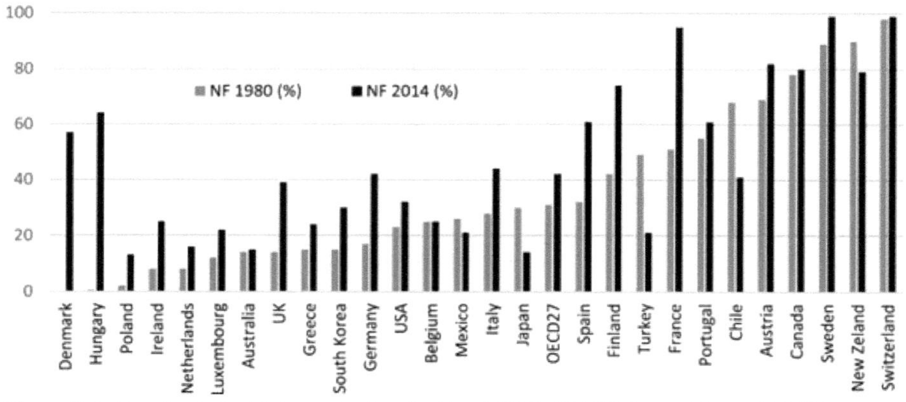

Figure 2.1. Non-fossil electricity production fraction (NF/T), 27 OECD countries and OECD average 1980 and 2014 [10].

noticeable in OECD countries, as can be seen in Fig. 2.1, where in some countries the production of energy by renewable means has already clearly surpassed the use of fossil resources. The evolution between 1980 and 2014 is also well known, with it being noticeable that some countries were already close to the goal of producing 100% of their electricity without resorting to fossil fuels in 1980.

The fight for decent work for all and economic growth is largely influenced by the policy followed by each country, depending on its natural resources, the intellectual capacities and skills of the population and the previously established economic dynamics [11]. The clear success of some countries in Asia, Eastern Europe and even in South America, shows that economic success and job creation depend a lot on the policies followed by the government of each country. The main problem associated with economic growth is consumption, which translates into garbage generation, pollution, in addition to the consumption of natural resources. The balance between these factors seems difficult to achieve, mainly because no country wants to jeopardize its economic growth, since this is a primordial comparative factor to show the degree of development of each country.

Employment generation is increasingly dependent on the industrial and service sectors. For these sectors to advance in a sustainable way it is necessary to have a high innovation for creating sustainable employment, which can provide as much as possible equal opportunities for the entire population. For this to be possible, technological advancement should occur at the same pace at which populations can evolve in terms of skills. The effort in training its population by each country should be constant and extremely tenacious, creating in the population the awareness that they need to be constantly learning, to accompany the evolution that our planet registers in social and technological terms. This effort by the populations should also be accompanied by an effort by governments to create the necessary infrastructure for this evolution. These conditions necessarily include providing conditions for the use of the most current means of communication, namely, the internet. In fact, only 54% of the world's population uses the internet, and in less developed countries, only 19% of the population has access to the internet [12]. Given that the internet is today a good that can be considered essential in terms of learning and communication for the development of the populations, the creation of the necessary infrastructure is an extremely important pillar in granting equal opportunities for evolution to the entire population. This is a crucial investment for developing society, after meeting the basic priorities described above.

Humanity has been prodigal in creating aspects that can generate inequalities. History tells us ancient stories, where social differences led to the glory of some, to the detriment of slavery of others. Fortunately, the barbaric acts reported in history seem to have diminished, at least in intensity, but much more needs to be done so that these inequalities dissipate even more, since, their elimination enters the field of utopia. Some aspects that lead to differences in the living conditions of populations have already been mentioned, such as gender, sexual orientation, salary, disability, race, ethnicity, religion, political orientation, and even age [13]. Any of these factors can create inequality of opportunities in society in which we are living, although complete freedom and equality of opportunities is proclaimed. Inequalities generate anguish and depression when one is a victim, and it is not rare at times when one degenerates into conflict. This inequality, when spurred on by those who take the lead, can generate serious social problems that are very difficult to solve. It has been common to create interest groups that, taking prominent positions, try to take advantage of them, digging even deeper into previously existing inequalities. The genesis of these inequalities often comes from the education that certain groups have, and others who did not have it because it was forbidden to them, or because they simply did not want to assimilate it. The UN reports that about 20% of the world's population has already experienced problems of discrimination in at least one of the points focused on human rights [13]. In fact, in 2022 there are still several countries where human rights are not respected, and the attribution of major events, such as the Football World Cup, to countries where there are strong doubts about respect for human rights is even being questioned. The construction of the stadiums that are at the base of the Football World Cup in 2022 resulted in 6500 deaths of workers involved in their construction, a situation that constitutes an indelible fact in the disrespect for safety in the work environment. In fact, it is the Qatari authorities themselves who reported that more than 15000 foreign workers died between 2011 and 2020, a period in which the construction works of the infrastructure necessary for the realization of that world event took place [14]. In 2022, women's right to freedom was also discussed in some Middle East countries, where the use of certain garments is mandatory, with no right of choice being granted, which constitutes a clear violation of the freedom of a given gender of the population. However, even in so-called moderate and democratic countries, human rights violations are constant, often induced by educational aspects that tend to become binding for these generations. This is an extremely difficult aspect to overcome, given that it is rooted in certain cultures, and that it only tends to be developed through disruptions between generations, through conflicts that should also be avoided.

Based on what was mentioned above, there is a perception in the minds of a large part of the population that there are more job opportunities in urban areas. Effectively, cases of isolationism can be considered almost an exception, as human beings tend to depend on others and seek the necessary support for their daily lives. The territorial organization is not always well-structured, and in more developed countries the decision-making centers and opportunities are better distributed territorially. In less developed countries, industrialization and the service sector tend to be concentrated in a smaller number of population centers, with these reaching a large dimension, regarding the population of each country. Areas in direct contact with the sea or with freshwater courses and lakes have always deserved the attention of the population, as they tend to have resources that are perceived as the most basic. The congregation of these factors with flat territorial parts, where agriculture became a source of livelihood or economic activity, was another decisive factor in the establishment and expansion of populations. However, industrialization and the services sector gave rise to an exponential growth of urban centers, to the point that, currently, these urban areas represent only about 3% of the earth's surface but are responsible for 60 to 80% of energy consumption and contribute with about 75% of the total CO_2 emissions [15]. Another major concern in the cities is the generation of waste. If, on one hand, municipal waste collection becomes easier in denser population environments, on the other hand, waste generation is much higher, and its management can become more complicated. Data released by the World Bank [16] point to the generation of 2.01 billion tons of municipal solid waste, which excludes solid waste produced by the industrial sector. It should also be noted that it is expected that around a third of this urban waste is not properly managed, considering public health rules and risks to human beings. World Bank estimates [16] also indicate that each person, considering the entire world population, generates between 0.11 kilograms and 4.54 kilograms of solid waste. Countries considered as developed are responsible for about 34% of all urban waste produced, that is, about 683 million tons. If these data are no longer encouraging, the prospects for 2050 are even worse. The World Bank [16] estimates that in 2050, 3.40 billion tons of solid waste will be produced, that is, an increase of more than 50% over the next 30 yr compared to today. An expectation was also made of the evolution of this production of solid waste by classification of countries according to their level of income. In fact, solid waste generation is expected to rise by more than 40% in middle income countries, while solid waste generation in higher income countries will rise by 19%. Concerning low-income countries, initially, there was a reduction in the generation of solid waste, but this trend was suddenly reversed, and there is now a rapid increase in the volume of solid waste generation. This increase predicts that by 2050, low-income countries will triple their current volume of solid waste production, which is extremely negative, given the current scenario. The expectation of an increase in the generation of solid waste is a negative fact, but the numbers are alarming. It is well-known that the care with the management of solid waste in developing countries is less careful, in general, than that which is carried out in developed countries, which makes the panorama even bleaker. In geographical terms, the areas of East Asia and the Pacific are those where the largest amount of solid waste is generated (23% of the total), while the regions of North Africa and the Middle East are those that produce the least solid waste, with only 6% of the total. However, some of these regions are also included in areas where the worst developments in solid waste production are expected, namely, Sub-Saharan Africa, South Asia, and the Middle East and North Africa. In these areas, it is expected that the volume of solid waste generated by 2050 will be more than two to three times higher than the values currently registered. In some of these regions, it is usual for garbage to be deposited in the open. Considering the usual temperatures in these areas, the degradation of waste under the effect of the sun and temperature occurs more quickly and dangerously, and can generate severe health problems for the population, in addition to generating serious environmental problems. As mentioned earlier, the management of urban solid waste is not carried out with the same care in all countries, with a significant difference being observed. In fact, while the highest income countries collect about 96% of the waste produced, in the lowest income countries, this percentage only reaches 39%, that is, just over a third of the waste produced follows a previously determined flow of management and treatment [16]. This fact, combined with the very sharp increase in waste production expected in this category of countries, should concern us all and requires that something needs to be done to change this trend, to increase the sustainability of society in general, and not just from higher-income countries. The typology of solid waste also varies with the income category of different countries. In fact, countries with lower incomes tend to produce more organic and green waste, from food and crop residues, to the detriment of recyclable dry waste (only 20%). Countries with higher incomes are more likely to produce dry waste, which is more easily recyclable, such as plastic, paper, cardboard, metal and glass (51%), most of them from packaging to the detriment of food and green waste (32%). In fact, only regions such as Europe, the United States and Central Asia generate less than 50% of organic waste, while all other regions generate values greater than 50% of this type of waste. The most developed countries with the highest income are also those that have the most developed technology to transform the disadvantages of producing solid waste, transforming into an advantage: the energy production [17,18]. However, developing countries have also made significant progress in the last two decades in the use of solid waste to generate various energy sources [19,20], which is to be welcomed. In fact, organic waste can also be partially converted into energy, through anaerobic digestion, as has been described by several researchers, who reported cases of success even in developing countries [21,22]. In more developed countries, care has been taken for more than a decade about how the cities of the future should be and, obviously, the collection, sorting and treatment of solid waste is an extremely important component to be taken care of. Associated with the Smart City concept, several studies have emerged that aim to create solutions capable of promoting adequate management of the collection, sorting, and treatment of solid waste in large population centers [23-25].

Sustainable consumption and production are other sensitive issues in terms of global sustainability. Consumption stimulates the economy and, therefore, is largely encouraged by the political class, as more consumption implies more economic activity, more employment, and the generation of more taxes, which allows satisfying some of the population's demands. The generation of more jobs agrees with the needs of social sustainability. However, more employment also means greater purchasing power, which can encourage greater consumption (in loop), but also greater waste generation, as well as harmful emissions. In fact, the economy should not only be stimulated by consumption, as more investment should be generated, in parallel with some savings. However, it is necessary to take care of production. If production is sustainable, safeguarding all the rules for saving the environment, increasing the life cycle of products, reducing the use of natural resources, and taking greater care in reusing or recycling products and materials, the harmful effects on the environment can be moderate, and production can be considered as acceptable in environmental terms. In this way, the increase in consumption is not largely impacting the environment. However, excessive consumption, on the other hand, usually leads to waste, shortening the life cycle of products, not conveniently valuing reuse, reconditioning, or proper recycling, which is greatly harmful in environmental terms. It is curious that this Sustainable Developing Global, when announced by the UN [26], does not clearly defend the consumption reduction, referring only that people should reduce waste and purchase sustainable products whenever possible. Political pressure on this issue is also undisguised, as governments perceive the reduction in consumption as a phenomenon that must necessarily affect the economy and employment. Perhaps it would not have to be that way if policies were to balance the economy, smoothly counteracting this effect. In any case, as consumers, we will have to be more vigilant regarding the origin of the products, the way they were produced, the materials used, the life cycle provided by the product and how they will be reused, reconditioned, recycled, or deposited in a landfill. This information should be included in the product's technical sheet, and consumers should receive adequate training to know how to interpret this information, thus allowing a more conscious choice between apparently similar products. Technological evolution and marketing pressure means that many consumers change their technological devices (mobile phones, computers, tablets, etc.) much faster than necessary. The result is obvious: the volume of electrical and electronic waste generated has recently grown by around 33%, and only around 20% of that waste is recycled [26]. Even in terms of recycling, the common consumer is usually unaware of the amount of energy that is needed to recycle an aluminum can, a glass bottle, or a plastic container. The beverage may be the same, but the environmental impact of packaging is not the same. Populations need to be confronted with real situations so that they can consciously make decisions. To this end, it is necessary to create educational programs that allow the population to internalize concepts that, even though they are basic, need to be properly assimilated, since, apparently, they do not interfere with people's daily lives, but they are certainly at the origin of increasingly more frequent and more severe climatic phenomena. It can thus be seen that there is still a long way to go, both on the side of producers and on the side of consumers, in the sense of producing and consuming in a more conscious way in terms of sustainability, and here three aspects of sustainability are involved: economical, environmental, and social [27].

The issue of climate is one of the most pressing issues, and one of the most mediatic, as it is shocking to witness the global warming that is occurring, the melting of glaciers, the rise in ocean levels with the possibility of some islands being completely submerged. Simultaneously, we are witnessing extreme weather phenomena, with fires of proportions never seen before, floods, extremely violent storms, etc. Slowly but consistently, these phenomena gradually worsen and multiply in frequency. Continuing to ignore the obvious is not a solution. It is urgent to rethink everything, from the exploitation of materials to their end of life, including the entire supply and processing chain. It is necessary to change the paradigms. Great deeds are done with humble gestures. But we should be aware that having a greener planet does not depend only on others: it really starts with ourselves. The target of not letting the global temperature increase by more than 1.5°C, as agreed in Paris in 2015 (COP 21), implied that there would be a reduction in the emission of greenhouse gases of 7.6% per year, starting in 2020 [28]. If the COVID-19 pandemic effects experienced in 2020 and 2021 slightly pushed back emissions due to the confinement periods experienced in almost global terms, in 2022, levels increased again. The agreement continues not to be fulfilled and the successive COP meetings (in 2022 the COP 27 occurred in Sharm El Sheik, Egypt) continue ending without any agreement. The problems with the climate do not seem to bother the political class, which is more concerned with the survival of its economy, and with its own political survival, than with the global well-being in the future.

Marine life is also a major concern, as the seas are the destination of much liquid and even solid waste. The production of plastics, after a few years of moderate growth, seems to have stagnated, registering a marginal decrease from 368 million tons in 2020 to 367 million tons in 2021 [29], with almost half of the world's production of plastics concentrated in Asia (49%). The images of the seacoast ~~s~~ and watercourses full of floating plastic packaging are familiar and frequent, especially in some Asian countries. Plastics partially make their way into the fish food chain in the form of microplastics. Microplastics have become abundant in aquatic ecosystems, because, fish have easy access to them due to human carelessness in managing plastic waste. Microplastics ingested by fish represent a source of toxicity for a wide range of fish species. Once ingested, microplastics tend to disperse through the body tissues, which can become contaminated with dangerous substances. Fish is a very common source of protein for humans, who are exposed by this route to microplastic particles ingested and absorbed by fish. Thus, several chronic illnesses can appear in humans not correlated before with the microplastics presence into the fish tissues [30]. But there are still many other reasons for concern for the aquatic environment. In less developed countries, watercourses are still one of the most used means for the disposal of liquid waste from industrial economic activity (chemical industry, textile industry, food industry, industrial slaughterhouses, etc.). Even in Europe, in the 20th century, it was common to install industrial units next to water courses for exactly this purpose: to easily drain the effluents generated by their economic activity. This fact shows that only in recent decades has there been greater awareness, and ~~also~~ greater control, over the pollution of water courses. At sea, the situation is even more serious and difficult to control. In addition to accidents with the extraction or transport of oil, related products, and chemicals, it is also common for a significant number of ships to wash their reservoirs on the high seas, contaminating the waters and, as a result, the marine species that inhabit them, which can later be fished, bringing substances that can be largely harmful in terms of health into the human food chain. Once again, economic interests and lack of environmental awareness overlap with environmental sustainability, requiring greater supervision so that the rules can be complied by all and the planet does not degrade at the current pace.

The degradation of the land we live on is also a serious concern. About 31% of the planet's land area is covered by forest, which is home to about 80% of the existing plants, animal species and insects [31]. However, land degradation and climate change are drastically reducing biodiversity. Effectively, about a fifth of the Earth's surface is degraded, thus affecting the lives and well-being of more than 3.2 billion people, and contributing to the extinction of many species. This degradation, in a closed cycle, also contributes to climate change. At the same time, it is known that the decline in biodiversity strongly undermines humanity's resilience to the adversities brought about by climate change.

Peace, justice, and robustness of institutions are more closely linked to social sustainability. However, this book is being written in wartime Europe. This situation not only degrades human life, showing complete disregard for freedom, the autonomy of populations and human life, but also contributes to a very considerable waste of resources. Here, the waste is multiple, and very considerable: (1) mainly, human lives are lost in a completely absurd way; (2) economic and material resources are wasted that add no value to humanity; (3) destroyed assets that need to be recovered; (4) territories are contaminated; (5) the normal flow of products produced in a given region is impeded; and (6) the worldwide economic and social environment deteriorates, in addition to many other factors. It can therefore be concluded that peace is a precious asset, as well as justice and the institutions that promote both peace and justice. Even so, in times of peace, the UN reports that there are around 100 people killed every day due to armed conflicts [32]. These institutions are run by people. When these people put personal interests or those of a restricted group ahead of the well-being of humanity, and when the power of certain institutions overrides the rationality that should govern the conduct of humanity, conflicts arise. Therefore, humans take the lives of other humans, this situation is inexplicable based on common sense. This demonstrates that institutions are not yet strong enough to effectively eliminate or lessen these conflicts, and that even justice is something that seems to be difficult to enforce, in light of the atrocities that are committed in times of war. In 2022, a great deal remains to be done in this field.

Cooperation between countries and institutions is the only possible way for all UN Sustainable Development Goals to be achieved [33]. A strong increase in existing partnerships could allow access by less developed countries to practices and technologies already successfully adopted by more developed countries. These practices, after being tested and validated, can easily be shared in databases that make it possible to minimize the effort of less developed countries to increase their sustainability policies and practices, thus contributing to a greener planet. The increasingly widespread use of the internet could allow less developed countries to access the "Technology Bank for Least Developed Countries", trying to shorten the existing differences in sustainable practices and regulations between more and less developed countries. Also, in terms of legislation and collaboration between countries, the creation of partnerships can help bring the practices of more and less developed countries closer together. It is necessary to emphasize that less developed countries should not omit or make the existing legislation in more developed countries more flexible, facilitating the migration of more polluting companies from developed countries to less developed countries, without a clear technological update to ensure that pollution does not pass from one country to another, continuing to pollute the same planet. Here, too, cooperation is extremely important, but it should put economic interests aside, which is not always respected by the political class.

It thus became clear that our planet faces colossal challenges, which have not seemed easy to resolve. Due to this, the UN developed a set of 17 principles that could make our planet more sustainable, which they designated as Sustainable Development Goals. These goals should function as a series of principles and guidelines that aim to eradicate some of the most severe problems that continue to plague humanity. However, these principles appear extremely difficult to implement, and every day our planet is degraded a little more, substantiating those political decisions to be taken, will need to be increasingly drastic, which makes them increasingly difficult to implement. The rate at which the change seems possible is not compatible with the urgency demanded by the environmental degradation that we observe. An analysis is then carried out of each of the Sustainable Development Goals, with a particular discussion on those that are more related to society and the environment, trying to dissect some weaknesses that the systems have in achieving these goals.

2.2 Sustainable development goals and green practices

Analyzing the description of the 17 principles announced by the United Nations as the main vectors for sustainable development (Fig. 2.2), it is easy to understand that, on one hand, these are extremely ambitious goals, and that some countries are much further away from achieving them than others. On the other hand, implicitly, it is an acknowledgment that human dignity is far from being a reality for everyone. Indeed, nobody believes that such ambitious goals will be achieved by the next generations. Between rhetoric and practice, there are many difficulties to overcome, and the path is not easy. But how are these goals in line with the practices of a greener world? This is what will be analyzed next.

Figure 2.2. UN Sustainable Development Goals [34].

Conceptually, everyone agrees in completely regarding the first objective: "End poverty in all its forms everywhere". However, will immensely wealthy people be willing to share their wealth so generously? At first instance, will every human being equally take advantage of all the opportunities available to them? We all know that this is not true, because the human mind has a number of ways of thinking towards life. However, opportunities are not the same for everyone, and there are many people who would like to seize an opportunity and fail to do so at all. The same question arises in terms of countries. When a polluting industry proposes to install a plant in a developing country, governments vacillate between reducing poverty in their country and hosting the polluting company, or not hosting that company and not generating wealth for their population. Thus, the decision is political, but also a business, which is rarely completely explained to the populations, such is the concern to find ways of economic development for the populations, even if this entails risks and the environmental standards that the others impose in their countries. Obviously, given the existing technological resources, the risks and levels of pollution are lower, but they are far from zero. If they were, there might be no reason to move the plant from its original location. On the other hand, one also has to consider less or non-polluting companies. Removing from this equation the ease of logistical operations of the operation of certain companies closer to their market, which is highly commendable, since it contributes to less pollution generated by the necessary transports, the choice of location of many companies is not carried out with the purpose of leveling the wealth generated by different countries, but to take economic advantage by paying lower wages for the same level of skills. As a negative aspect, it must be considered that these companies try to get a competitive advantage-over others by choosing a country that offers lower operating costs. As a positive aspect, note that the generation of employment reduces poverty, according to the objective defined by the UN and may even provide conditions for its employees to also become customers of the product they manufacture, increasing satisfaction. from both parties.

It can also be mentioned that the second objective of the UN, "End hunger, achieve food security and improved nutrition and promote sustainable agriculture", is closely linked with everything mentioned above. Fortunately, in some developed countries, companies have a very compelling social concern, leading to the investment in the generation of employment and wealth being evenly distributed across the different regions of a given country, thus ensuring the fixation of populations to their roots. On the other hand, we are witnessing the stampede of large groups of people who, given the lack of opportunities, seek a better situation at any price, even if they put their own lives at risk. There are too many cases in which countries do not properly care for their populations, not for lack of natural resources, but only for political reasons. As for the promotion of sustainable agriculture, it is an aspect that is perfectly in line with the principles of greener production, through the reduction or elimination of the use of chemicals harmful to the environment. These first two goals are in line with the fourth goal, "Ensure inclusive and equitable opportunities for all", which is not directly related to the environment, but is clearly linked to the first two goals. When the economy is divided between developed and developing or less developed countries, it is clearly assumed that equal opportunities are far from a reality. Even within so-called developed countries, equality of opportunity is a utopia, both in so-called democratic and totalitarian regimes. Unfortunately, given the politics that are practised and the nature of the human being in general, many of these principles can be considered utopian. It works as a tenuous line of thought, to which some attention is sometimes given, but which is far from constituting an ingrained practice with short or medium-term goals.

The third goal, "Ensure healthy lives and promote well-being for all at all ages", depends largely on ourselves, through the lifestyle we choose, and the way we eat, among other factors. In any case, it should also be mentioned that poverty eliminates it completely, or almost, the possibility of choosing the lifestyle and food that certain people want to have, strongly limiting the options and leading human beings to conditions in which health is not safeguarded at all. Human health also depends largely on the surrounding environment, and highly polluted urban or industrial environments do not guarantee this principle, so there are very strong constraints to overcome.

The sixth and seventh goals, "Ensure availability and sustainable management of water and sanitation for all" and "Ensure access to affordable, reliable, sustainable and modern energy for all", are closely linked with the environmental objectives, depending largely on the first, second, and fourth objectives. Due to extreme weather events, drinking water is already a serious problem in many areas of the planet, which contrasts with severe flooding in other areas [35]. Agriculture is both a cause and a victim of these climate changes. The work in adopting adequate irrigation policies is slow to become a generalized practice, or even mandatory, leading to water scarcity. On the other hand, droughts, and floods severely damage crops. In this economic sector, there is still a lot to be done, and automation can play a critical and crucial role in conditioning irrigation systems, contributing to a more rational consumption of water. Food education is also a crucial point to be developed, as the exaggerated consumption of meat leads to the need for adequate food for these farm animals, which implies large amounts of water. Access to energy mainly depends on generation and distribution policies. Costs are strongly influenced by economic factors that are not always clear, called market prices, which were recently strongly influenced by political factors, mainly on the European continent. However, the constraints created by political factors could provide an incentive for politicians themselves to make decisions to encourage the adoption of greater amounts of renewable energy.

The eighth goal, "Promote sustained, inclusive and sustainable economic growth, full and productive employment and decent work for all", is in line with the distribution of wealth and the correct establishment of companies that, in addition to profit, also aim to provide the comfort conditions necessary for populations. However, the installation of these companies should not be allowed at any price, as mentioned above.

The ninth goal, "Build resilient infrastructure, promote inclusive and sustainable industrialization and foster innovation" is directly related to the theme covered in this work: to produce, but in a more sustainable way, both economically, socially, and environmentally. If most of the goals are centered on human dignity, this one is also on the same line, but considering that we should care for the planet where we live. This needs to be a permanent concern to reverse the trend of harm created so far.

The 10th goal, "Reduce inequality within and among countries" is one of the simplest to outline, but perhaps the most utopian one. The geographic and geopolitical positioning, different access to natural resources, the climatic conditions, the political management, and the inequality existing at the time, make this objective impossible to achieve. It is curious to realize that some countries will tend to disappear, as they remain submerged, having contributed in an extremely small way to this. This remained clear during COP 27 in Egypt, in November 2022 [36].

The 11th goal, "Make cities and human settlements inclusive, safe, resilient and sustainable", touches directly on this subject : it is impossible to guarantee a dignified human existence in extremely polluted environments and, for that, it is necessary to make an extra effort towards making everything around us greener, in other words, to drastically reduce pollution levels. With high levels of pollution, it is impossible to guarantee the safety of populations, and if consumption continues to be exacerbated, nature cannot renew itself at the same pace, creating serious sustainability problems. This also refers to the 12th goal, "Ensure sustainable consumption and production patterns". The key to sustainability lies largely in reducing consumption. However, if the economic system is not prepared for this, it will collide with the first goal, namely with the eighth goal, which in turn is related to the first three. The balance between these factors is something very difficult to achieve, but for that, it is necessary that there is an awareness among the population, especially those with greater purchasing power, to make consumption more rational and sustainable. The transition, to occur would have to be carried out very smoothly over time, allowing the economies to adapt to a lower consumption rate. However, it should be noted that education for sustainability is something that can take several generations to achieve and with minimally acceptable results.

The 13th goal, "Take urgent action to combat climate change and its impacts", is the key point. The degradation of climatic conditions is visible to all, with some suffering more than others. Despite the political discourse, it is not easy to reverse all the wrong paths taken so far, in an expeditious way. Economic power trumps political power, and sufficiently effective actions are not taken to reduce emissions. Will it take a global catastrophe to rethink the way we treat the environment? Politically, tenuous steps have been taken to change the course of events. The automotive industry is changing paradigms in terms of energy sources. The way in which electricity is generated is also being reconsidered, eliminating the most polluting forms, namely, using coal. In individual terms, if the consumerist fervor persists and the desire to travel overcomes the awareness that the means of transport, for the most part, are extremely polluting, the planet will hardly be able to converge towards this goal.

The 14th goal, "Conserve and sustainably use the oceans, seas and marine resources for sustainable development", is closely linked to the emissions of polluting effluents released into waterways, which invariably end up in the sea. In the 20th century, in some developing countries, it was customary for certain polluting industries such as textile dyeing, industrial slaughterhouses, metallic surfaces electroplating, among others, to be set up next to rivers, so that the effluents produced were quickly drained by the current of the waters. Steel manufacturers are often located near rivers and lakes, given their need to collect water, also facilitating its drainage after use. In countries with greater environmental concerns, water is collected, used, treated/purified, and returned to the collection site, without affecting the environment. There is technology for this, so the only thing missing is environmental awareness, or legislation that obliges the technology to be used. However, it is not just the industrial sector that is responsible for this pollution. The use of harmful chemicals in agriculture, such as some fertilizers and pesticides, ends their cycle in oceans after leaching of the land by rains. Nevertheless, there are solutions that produce the same effect of fertilization and protection against pests, without their leaching contaminating the water courses. The use of these kinds of solutions, normally more expensive, could create asymmetries in terms of competitiveness, hence global legislation is needed to make this the only possible way. For this, a global agreement is needed, which could very well occur in the Summits called Climate Changes Conferences, such as the recent COP 27 promoted by the United Nations. However, the larger the group, the more difficult consensus becomes. The asymmetry of conditions observed among many countries would hardly allow an approximation capable of generating disruptive ideas with what are common practices nowadays.

The 15th goal, "Protect, restore and promote sustainable use of terrestrial ecosystems, sustainably manage forests, combat desertification, and halt and reverse land degradation and halt and biodiversity loss", is closely and mainly linked to the migratory flow of populations towards urban areas and the savage felling of large forest reserves. This objective can only be achieved with very assertive policies for settling populations through an adequate implementation of businesses capable of guaranteeing the subsistence of families and creating wealth in regions prone to desertification. Migratory flows from North Africa to Europe and from South America to the United States of America are problems that date back to the last century, but which remain unresolved. Companies continue to install themselves preferentially in Europe, the United States of America and China, demanding more and more resources in terms of manpower, which leads to the flow of people from other continents to these regions. A better distribution of the establishment of businesses would certainly generate greater social balance, avoiding the desertification of some regions, and the over occupation of others. Regarding deforestation, human greed, and extreme weather phenomena, which generate fires of alarming proportions and the burning of huge areas of forest, are increasingly frequent problems that can only be mitigated through adequate policies for protecting essential forest areas, and to more courageous environmental policies, capable of reversing the current trend.

The 16th goal, "Promote peaceful and inclusive societies for sustainable development, provide access to justice for all and build effective, accountable and inclusive institutions at all levels", as well as the 17th and final goal "Strengthen the means of implementation and revitalize the Global Partnership for Sustainable Development", essentially concern social and political problems, some of which may also appear from economic issues and policies followed by each country. However, it can be considered that it is outside the scope of this work, so it is not important to dissect it here.

From the detailed description of the 17 UN Sustainable Development Goals, it is easy to observe that a good part is directly related to the environment, and that, if they were achieved, we would have a much greener and more sustainable environment. Most of the goals become utopian, given the installed economic interests, the geographic and geopolitical conditions, the political regime, and orientations followed by each country or region, the natural climatic conditions, the natural resources of each country, the way society is organized, the educational level, as well as the different degrees of ambition and way of being in each individual's life. Reaching the goals meant that the world had been able to overcome all these problems and overwhelm all these differences. Then, it can be considered that they are unattainable goals at all. However, the principle that should have guided the design of these goals is to make society think and do something every day that allows us to get a little closer to these goals. But, at the rate at which we are evolving, will we still have time? In global terms, what prospects are we creating for our children and grandchildren? Could it be that the excessive ambition in the present is taking away the lucidity of understanding what the future could be? These are questions that any conscious human being cannot ignore.

References

[1] UN. (2022). Zero Hunger: Why it Matter", URL: https://www.un.org/sustainabledevelopment/wp-content/uploads/2016/08/2_Why-It-Matters-2020.pdf (Accessed on November 24th, 2022).

[2] UN. (2022). "Good health and well-being: Why it Matter", URL: https://www.un.org/sustainabledevelopment/wp-content/uploads/2017/03/3_Why-It-Matters-2020.pdf (Accessed on November 24th, 2022).

[3] UN. (2022). Zero Poverty: Why it Matter, URL: https://www.un.org/sustainabledevelopment/wp-content/uploads/2016/08/1_Why-It-Matters-2020.pdf (Accessed on November 24th, 2022).

[4] UN. (2022). Quality education: Why it Matter, URL: https://www.un.org/sustainabledevelopment/wp-content/uploads/2017/02/4_Why-It-Matters-2020.pdf (Accessed on November 25th, 2022).

[5] UN. (2022). Gender equality: Why it Matter, URL: https://www.un.org/sustainabledevelopment/wp-content/uploads/2016/08/5_Why-It-Matters-2020.pdf (Accessed on November 25th, 2022).

[6] UN. (2022). Clean water and sanitation: Why it Matter, URL: https://www.un.org/sustainabledevelopment/wp-content/uploads/2016/08/6_Why-It-Matters-2020.pdf (Accessed on November 25th, 2022).

[7] UN. (2022). Affordable and clean energy: Why it Matter, URL: https://www.un.org/sustainabledevelopment/wp-content/uploads/2016/08/7_Why-It-Matters-2020.pdf (Accessed on November 25th, 2022).

[8] Thombs, R. P. (2022). The asymmetric effects of fossil fuel dependency on the carbon intensity of well-being: A U.S. state-level analysis, 1999–2017, Global Environmental Change, 77: 102605. doi: 10.1016/j.gloenvcha.2022.102605.

[9] Nerlinger, M. and Utz, S. (2022). The impact of the Russia-Ukraine conflict on energy firms: A capital market perspective. Finance Research Letters, 50: 103243. doi: 10.1016/j.frl.2022.103243.

[10] Morten, J., Skonhoft, D. A., Christensen, M. Q. and Ødegaard, E. T. (2019). Does economic growth eat up environmental improvements? Electricity production and fossil fuel emission in OECD countries 1980–2014, Energy Policy, 125: 103–109. doi: 10.1016/j.enpol.2018.10.051.

[11] UN. (2022). Decent work and economy growth: Why it Matter, URL: https://www.un.org/sustainabledevelopment/wp-content/uploads/2016/08/8_Why-It-Matters-2020.pdf (Accessed on November 25th, 2022).

[12] UN. (2022). Industry, information and infraestruture: Why it Matter, URL: https://www.un.org/sustainabledevelopment/wp-content/uploads/2019/07/9_Why-It-Matters-2020.pdf (Accessed on November 25th, 2022).

[13] UN. (2022). Reduced inequalities: Why it Matter, URL: https://www.un.org/sustainabledevelopment/wp-content/uploads/2018/01/10_Why-It-Matters-2020.pdf (Accessed on November 25th, 2022).

[14] Walter, J. D. and Ford, M. (2022). DW - Deutsche Welle, Fact check: How many people died for the Qatar World Cup?, URL: https://www.dw.com/en/fact-check-how-many-people-have-died-for-the-qatar-world-cup/a-63763713 (Accessed on November 25th, 2022).

[15] UN. (2022). Sustainable cities: Why it Matter, URL: https://www.un.org/sustainabledevelopment/wp-content/uploads/2019/07/11_Why-It-Matters-2020.pdf (Accessed on November 28th, 2022).

[16] World Bank. (2022). Trends in Solid Waste Management, URL: https://datatopics.worldbank.org/what-a-waste/trends_in_solid_waste_management.html (Accessed on November 28th, 2022).

[17] Shahab, A. V., Srivastava, V. K., Mohanty, S. S. and Varjani, S. (2021). Municipal solid waste as a sustainable resource for energy production: State-of-the-art review. Journal of Environmental Chemical Engineering, 9(4): 105717. doi: 10.1016/j.jece.2021.105717.

[18] Koroneos, C. J. and Nanaki, E. A. (2012). Integrated solid waste management and energy production—a life cycle assessment approach: the case study of the city of Thessaloniki. Journal of Cleaner Production, 27,: 141–150. doi: 10.1016/j.jclepro.2012.01.010.

[19] Kefalew, T. and Lami, M. (2021). Biogas and bio-fertilizer production potential of abattoir waste: implication in sustainable waste management in Shashemene City, Ethiopia, Heliyon, 7(11): e08293. doi: 10.1016/j.heliyon.2021.e08293.

[20] Mbuligwe, S. E. and Kassenga, G. R. (2004). Feasibility and strategies for anaerobic digestion of solid waste for energy production in Dar es Salaam city, Tanzania, Resources, Conservation and Recycling, 42(2): 183–203. doi: 10.1016/j.resconrec.2004.02.010.

[21] Moktadir, M. A. and Rahman, M. M. (2022). Energy production from leather solid wastes by anaerobic digestion: A critical review, Renewable and Sustainable Energy Reviews, 161: 112378. doi: 10.1016/j.rser.2022.112378.

[22] Papa, G., Sciarria, T. P., Carrara, A., Scaglia, B., D'Imporzano, G. and Adani, F. (2020). Implementing polyhydroxyalkanoates production to anaerobic digestion of organic fraction of municipal solid waste to diversify products and increase total energy recovery. Bioresource Technology, 318: 124270. doi: 10.1016/j.biortech.2020.124270.

[23] Abuga, D. and Raghava, N. S. (2021). Real-time smart garbage bin mechanism for solid waste management in smart cities, Sustainable Cities and Society, 75: 103347. doi: 10.1016/j.scs.2021.103347.

[24] Esmaeilian, B., Wang, B., Lewis, K., Duarte, F., Ratti, C. and Behdad, S. (2018). The future of waste management in smart and sustainable cities: A review and concept paper, Waste Management, 81: 177–195. doi: 10.1016/j.wasman.2018.09.047.

[25] Shah, P. K., Anagnostopoulos, T., Zaslavsky, A. and Behdad, S. (2018). A stochastic optimization framework for planning of waste collection and value recovery operations in smart and sustainable cities, Waste Management, 78: 104–114. doi: 10.1016/j.wasman.2018.05.019.

[26] UN. (2022). Responsible consumption & production: Why it Matter, URL: https://www.un.org/sustainabledevelopment/wp-content/uploads/2019/07/12_Why-It-Matters-2020.pdf (Accessed on November 28th, 2022).

[27] Silva, F. J. G., Kirytopoulos, K., Ferreira, L. P., Sá, J. C., Santos, G. and Nogueira, M. C. C. (2022). The three pillars of sustainability and agile project management: How do they influence each other. Corporate Social Responsibility and Environmental Management, 29: 1495–1512. doi: 10.1002/csr.2287.

[28] UN. (2022). Responsible consumption & production: Why it Matter, URL: https://www.un.org/sustainabledevelopment/wp-content/uploads/2019/07/13_Why-It-Matters-2020.pdf (Accessed on November 29th, 2022).

[29] PlasticsEurope. 2022). Plastics e the Facts 2021. URL: https://plasticseurope.org/knowledge-hub/plastics-the-facts-2021/(Accessed on November 29th, 2022).

[30] Bhuyan, S. (2022). Effects of Microplastics on Fish and in Human Health, Frontiers in Environmental Sciences, 10, 827289. doi: 10.3389/fenvs.2022.827289.

[31] UN. (2022). Life on land: Why it Matter, URL: https://www.un.org/sustainabledevelopment/wp-content/uploads/2019/07/15_Why-It-Matters-2020.pdf (Accessed on November 29th, 2022).

[32] UN. (2022). Peace, justice and strong institutions: Why it Matter, URL: https://www.un.org/sustainabledevelopment/wp-content/uploads/2019/07/16_Why-It-Matters-2020.pdf (Accessed on November 29th, 2022).

[33] UN. (2022). Partnership: Why it Matter, URL: https://www.un.org/sustainabledevelopment/wp-content/uploads/2019/07/17_Why-It-Matters-2020.pdf (Accessed on November 30th, 2022).

[34] United Nations. UN Sustainable Development Goals, retrieved from: https://www.un.org/sustainabledevelopment/blog/2015/12/sustainable-development-goals-kick-off-with-start-of-new-year/, accessed on November 7th, 2022.

[35] Silva, F. J. G. and Gouveia, R. M. (2020). Sustainable production cases. pp. 281–373. *In*: Silva, F. J. G. and Gouveia, R. M. (Eds.). Cleaner Production – Toward a better future. Springer Nature Switzerland, Cham, Switzerland, 2020. doi: 10.1007/978-3-030-23165-1_8.

[36] United Nations Foundation. United Nations Climate Change Conference (COP27). retrieved from: https://unfoundation.org/cop27/?gclid=CjwKCAiAvK2bBhB8EiwAZUbP1Dy8A-C3io8EeLYDLjFee96676udGsnwhfz3XUtgRkQerTufZHLR7BoCIIgQAvD_BwE, accessed on November 9th, 2022.

CHAPTER 3

How Can Automation Make the World Greener

3.1 Contextualization

The consumption of natural resources and climate changes due to pollution are some of the main concerns regarding the desire to care for the environment and make the planet sustainable. Approaches to achieve results at this level can take a large number of forms. The one that will be dealt with in this work will be the help that automation can bring in the sense of reducing the negative effects of current processes and procedures on the environment.

Although some of the leisure activities also include various forms of pollution and even strong ecological footprints, such as traveling, for example, most of the aggressions caused to the environment come from economic activities and commuting between home and work. In fact, even when we are not working, we are often consuming, which implies that there is a production activity behind it, even if it is just the production of electricity.

Whether products are obtained through farming or processed or produced industrially, they have an impact on the environment. Automation and robotics have played a key role in (1) taking workers away from hard work, (2) performing dangerous tasks for humans, (3) performing repetitive tasks, among many others. Since it is necessary to ensure the continuity of human existence, man does not have to live only for work, nor should work be seen as a fatality for man. Indeed, mainly in high-income countries, new currents of thought argue that if the population worked less time, this would be beneficial for the environment, since the level of pollution and consumption of resources would be smaller [1]. This seems to be a convenient argument to reduce working hours and increase free time for inhabitants of developed countries, but the discussion around this issue is far from being confined to this aspect and may even be inconclusive. By this way, maybe social sustainability could be strengthened, reducing unemployment, and offering better conditions for a more balanced family life. However, some questions remain to be answered, namely: (1) working fewer hours, would the production volume be sufficient to produce everything that is required by the market? (2) would the additional free time gained eventually be used to increase consumption? (3) would some leisure activities practised in the additional free time eventually be just as or more harmful to the environment than the economic activity suppressed? The first topic takes one to the need for automation. In fact, automation solves numerous problems in terms of production and, if properly thought out, processes can be optimized through automation, including robotization. However, automation has also been the subject of great controversy in terms of generating unemployment. This topic is equally controversial, as developed countries normally have difficulty finding operators for processes that require greater expenditure of energy and physical activity. In addition, repetitive tasks cause professional illnesses and high mental stress, which is why automation is very crucial to combat operators' lack of interest in various functions and to avoid professional illnesses. Thus, the reasoning presented by Gomez-Baggethun [1] is interesting from a social perspective, but much less interesting from an environmental sustainability perspective. In fact, as the same author concludes, it is still necessary to investigate whether a smaller volume of work is enough to provide all the goods that society needs, in the way it is instituted, leaving the environmental aspect completely open, since, there is another variable not included in that study and which has a strong influence on the answer to this question: how will consumption evolve in the coming decades? If one looks only at the trends dictated by statistics, consumption will continue to grow, barring any more serious economic crisis. Given the environmental perspective, and clearly assuming that current consumption is unsustainable, consumption should decrease. However, to do so, it is necessary to create an environmental awareness that currently does not exist in most humans, and which will always tend to be contradicted by marketing.

Despite all widely recognized benefits of automation and robotics, it is necessary to remember that the means used for the production or transformation of goods use up resources, whether in the form of electric current or compressed air, the latter also consuming electricity, being less efficient than direct electric drive [2]. Zhang et al. [3] clearly stated that the replacement of human labor by robots and systems based on more conventional automation entails an extra consumption of electrical energy, which is harmful to the environment, for the reasons already mentioned in earlier chapters. However, if it is a fact that automation and robotics lead to increased pollution through the need to generate electricity, it is no longer so safe to say that productivity gains do not compensate for the resources consumed in powering these systems. Liang et al. [4] suggested that there is a clear focus for future studies toward ~~s~~ increasing productivity, while simultaneously trying to reduce energy consumption, using more efficient technologies. However, this claim is criticized by some scholars who claim that energy saving and technological progress are not compatible with the challenges posed by sustainable development [5]. In fact, it is unsafe to say that technological progress in terms of automation and robotization of processes has brought significant improvements in environmental terms, as there are clear signs of continued degradation of the environment and climate change, despite the existence of many other technologies that help in the most diverse aspects that this degradation is not even more expressive. Furthermore, a study carried out in 2007 in Sweden by Brännlund et al. [6] concluded that an improvement in energy efficiency of around 20% due to the development of new technologies translated into an increase in CO_2

emissions of around 5%. These results prove that even improving energy efficiency, there is a direct benefit to the environment, through a reduction in pollution. In fact, to comply with certain technological developments to increase energy efficiency, other production requirements are necessary often implying an increase in costs and the means used, which can impact the environment in a different way than intended. It should also be noted that if the product draws attention for becoming more energy efficient, consumers may increase demand for the product, which will translate into several negative factors for the environment: (1) more resources consumed producing a larger quantity of these products; (2) disposal of products that could still have a considerably useful life without being recycled or sent to a landfill; (3) possible use of more environmentally harmful means in the manufacture of these more efficient products. Thus, the simple exchange of a less efficient product for another energy efficient one, may not be the best option in the short or medium term for the environment. Conducting a careful balance of the benefits and threats to the environment is something that is only able to be performed by those who have an adequate level of skills and training in this area, that is, an almost insignificant amount of the world's population. Hence, in addition to the need for a very significant increase in training and awareness in the environmental area, additional information is also needed on the part of manufacturers and suppliers of all products (because they all have an environmental impact), in order to clarify through a simple and objective approach to the population, allowing the consumer to make a correct selection of the product that presents the least impact on the environment within a certain range. If producers had to include this classification, in the form of a perfectly understandable index or color, as is already mandatory in certain countries regarding the energy efficiency of household appliances, perhaps this would have a more relevant impact on manufacturers during the product development process. Effectively, knowing that the consumer could scrutinize the product based on simple and intuitive labeling, the manufacturer would feel the obligation to become more competitive in this matter as well, placing extra care in product development to obtain the best possible classification, and thus to be able to collect a greater preference from the consumer. It can thus be seen that automation is something critical in reducing the environmental impact that each product can assume, but other factors that need to be equally considered, with a view to making options that can be effectively beneficial to the environment.

In fact, automation and robotics can induce contradictory effects in what is normally called green production. On the one hand, automation and robotics can significantly increase the energy efficiency of processes, which translates into a clear benefit for the environment. On the other hand, there are factors that negatively and clearly affect environmental sustainability, such as: (a) investment is usually much higher, which reflects the level of resources that are needed to adopt this type of production; (b) if there is a decrease in the cost of the product, even considering the need to return the investment, this could reduce the price of the product on the market, increasing consumption, which goes against the intended effect of saving resources, reducing consumption; (c) if the price is cheaper and the market buys in larger quantities, this translates in a higher level of pollution generated, which presents a serious harmful effect for the environment. Despite being just three negative factors affecting the environment, they have a sufficiently negative influence to converge on the opinion that the benefit brought in terms of energy efficiency can revert to an even more negative balance for the environment. Therefore, automation and robotics bring clear advantages for economic and social sustainability, but it is not clear that they bring clear advantages for environmental sustainability. In fact, there is no point in being more efficient and saving resources per unit of product in the production, if consumption increases significantly and it is necessary to produce many more units of that product. The same was found by Binswanger [7], who found that when the cost of fuel was lower, people tended to travel longer distances. In an ideal situation, the larger scale at which a product is produced should be offset by a decrease in the level of pollution induced by this increased volume of production due to the adoption of technologies considered green. Thus, automation and robotics, together with the adoption of clean technologies, would lead to a level of emissions that, even by increasing the production volume, would positively offset the negative effect of the increase in the consumption of natural resources and generated emissions. Although in certain cases this situation seems possible, especially when very polluting production processes and raw materials are being used, this situation will be somewhat utopian in products and processes that have already undergone technological updates in the last two decades.

Environmental sustainability and social sustainability may not benefit in parallel in the same way through the adoption of systems based on automation and robotics. In fact, Nazareno & Schiff [8] stated that automation and robotics can contribute to a reduction in workers' stress, but also bring factors that negatively affect these workers, especially when analyzed from the perspective of complementarity of the work performed. Having analyzed several aspects of workers' well-being, namely in terms of stress, health, insecurity and job satisfaction, in a period between 2002 and 2018, the study has claimed that there was a decrease in the level of stress that proved to be more accentuated in the first years of the study and that has been decreasing since then, but, on the other hand, workers feel increased risks in terms of insecurity in relation to their job, there has been a decrease in the level of health of workers, and the job satisfaction also declined. However, it should be noted that the study did not incorporate any initiative to explain to workers whether the technological transformation also improved environmental sustainability, but it can easily be deduced that such awareness would not change their way of facing a possible loss of their job. Effectively, in this case, there would be a total opposition between environmental and social sustainability, since to minimize the effects of the environmental impact, a job would be put at risk, that is, create problems of social sustainability. These results clearly indicate that there must be additional concerns when adopting production systems based on automation and robotics. In fact, Araújo et al. [8] implemented an automatic system with mainly economic purposes, cutting two jobs, but there was a parallel concern to increase the skills of these workers in order to adapt them to new functions, which can perfectly pass through the programming of new automatic systems, or even for the maintenance of these same systems. Thus, it appears that there must be two essential concerns on the part of employers: explaining very well what are the reasons that underlie the adoption of automation and robotics, and education for environmental issues, when these are effectively what are in the basis of the decision. Obviously, when automation and robotics are adopted essentially with a view to making the product more competitive or the process more flexible, radical measures should probably be taken to eliminate total or almost total human intervention, whenever possible, to avoid the situations referred to in the study [9]. It is also of common knowledge that automation and robotics are adopted only with strictly environmental objectives when legislative requirements require it and, even so, any company will also intend to take advantage of this investment in technology to increase the competitiveness of the product. Therefore, it will obviously be difficult to combine environmental sustainability with social sustainability through the adoption of processes based on automation and robotics, especially if complementary jobs are carried out by workers. In fact, and mainly for economic reasons, Manyika et al. [10] claimed in a study published in 2017 that automation and robotics put at risk between 3 and 14% of the current workforce, while Nedelkoska and Quintini [11] reported that about 44% of jobs in the OECD face medium or high risk of extinction due to possible adoption of automation and/or robotics.

Despite the criticisms that may hover around automation and robotics implementation, their contribution to greener production processes is a fact, and it is necessary to conveniently assess where they best fit, taking into account the desired improvement of processes in terms of sustainability, which takes on different names, such as "environmentally driven innovation", "ecological innovation" or "green innovation", but which essentially aims to reduce the negative effects that manufacturing processes normally inflict on the environment. Gan et al. [12] described as extremely positive the application of robots in industrial processes in order to implement more environmentally friendly processes. However, at the same time, a description of the implications is made, which, according to the author, are positive, however neglecting some details. Effectively, together with the environmental benefits brought by the implementation of automation and robotics, cost reduction is always referred to, which, in turn, leads to more competitive prices for products, making them accessible to a larger share of the population, i.e., it stimulates consumption, with the negative consequences already mentioned. In addition, the same authors stated that the workforce decreases, a fact that contradicts some of the Sustainable Development Goals established by the UN, creating unemployment. These same authors also claimed that the use of robots makes processes simpler, making the use of more qualified labor unnecessary, translating into a decrease in wages paid to operators, which has two direct negative implications: (a) increases the competitiveness of the product, once again stimulating consumption and, therefore, increasing production, with a greater volume of emissions and greater consumption of natural resources; (b) it promotes the unemployment of more qualified people and encourages the employment of less qualified people, not being in accordance with some of the Sustainable Development Goals also defined by the UN. Thus, even strongly defending the application of automation and robotics in industrial processes, it is impossible to hide the negative implications that this may have even for the environment itself, in addition to all the social implications that it may also entail in terms of social sustainability. Based on the above, it is important to note that automation and robotics bring clear benefits in terms of economic sustainability, but in terms of environmental and social sustainability, the positive effect is noticeable, but only in certain sectors of activity. Thus, it is necessary to have a very comprehensive view of all the implications and carry out an analysis as correct as possible, so that the benefits in all aspects of sustainability can be a reality. In the work by Gan et al. [12], the perspective of savings on wages is presented differently, in which the income achieved by the company through lower wages paid would be reinvested in more environmentally friendly technologies, thus fueling a positive cycle, with clear benefits for the environment. However, if this can and should be true, we all know that companies essentially intend to present increasingly positive results for their investors, and that profit is the main focus of any company and investor. Thus, and although reinvestment is highly desirable for the proposed purpose, the balance between benefits and negative effects needs to be weighed. In any case, the salary reduction cannot be taken for granted. In fact, automation and robotics promote a division of the necessary skills into two groups: (a) one of the operators of automated systems, who now need fewer skills because they only monitor the progress of the process or promote the setup of the equipment; (b) others with much more specialized operators, who should have skills in terms of programming and problem solving, as well as others dedicated to the maintenance of automated equipment. If the first group is normally salaried with lower levels than a minimally specialized operator, the same is not true about the second group, which demands wages much more compatible with their level of skills and, given the scarcity of these types of skills in the market, given the strong increase in demand, they usually require very compatible wages. Analyzing literature, it is possible to state that there is a high heterogeneity in the effect that automation and robotics produce in the labor market. Bonfiglioli et al. [13] stated that robotics increases the placement of workers with higher skills. However, Acemoglu and Pascual [14] claimed that there is a greater tendency to a greater differentiation of wages in periods in which the use of automatic systems shows a strong trend of implementation. On the other hand, the existing heterogeneity in the application of automation and robotics can also differently affect inequalities in terms of wages paid to workers. Acemoglou and Pascual [15] stated that automation requiring highly skilled workers can negatively affect inequality, while automation that requires low-skilled workers positively influences inequality.

The implementation of automation and robotics cannot be seen as positive or negative in any type of industry or region. In fact, automation and robotics can be a promising option to remove workers from more densely polluted areas, thus contributing to the non-degradation of their health status. Indeed, despite all efforts normally made to minimize environments that are polluted, very humid or subject to extreme temperatures, certain processes require these conditions. Not being the most recommended conditions for human beings, but being necessary to perform tasks under these conditions, the use of automation and robotics can contribute to social sustainability, while environmental sustainability cannot be improved by changing operating conditions. The same applies in cities or areas where there are strong constraints in terms of pollution. It has also been observed that there is a greater tendency to implement automation and robotics in companies that devote generous budgets to R&D, or that develop and manufacture high-tech equipment [16].

The automation of processes, both to increase economic and environmental sustainability, has seen many developments in various economic sectors, although at different paces and with different strategies. Next, some approaches carried out using automation and robotics will be described both to increase environmental sustainability, even if contributing equally to economic sustainability, in sectors such as agriculture and livestock, civil construction, industry in the general, as well as the services sector.

3.2 Automation in agriculture and other green spaces

Agriculture, farming and even some service sectors with large green areas, such as resorts and golf courses, normally present non-negligible problems to environmental sustainability. In the area of agriculture, there are problems due to the high consumption of water, as well as the use of chemical products to eradicate pests and contributing to an increase in productivity, such as pesticides and fertilizers. In the field of farming and livestock, there are considerable problems with animal droppings, water consumption and wastewater generation. In resorts and golf courses, to keep the areas properly green and as free as possible from weeds, large amounts of water are also consumed, pesticides and fertilizers are also used. These problems have been identified for a long time, and climate change has sounded alarm bells for these sectors. The number of studies around these problems has been considerable, and here also the help of automation can mean an enormous added value in the sense of making the practices in these sectors much more environmentally friendly.

In agriculture, the chemicals used are also responsible for the death of around 300,000 people a year due to poisoning. Additionally, the harmful effects of some chemical products on human health are also known, namely in the development of some types of cancer. To increase sustainability and drastically reducing the use of chemicals in agriculture, organic production is one of the ways identified as the most viable, although perhaps not the most productive.

Biotechnological agriculture had fervent supporters, but also many enemies, who built a negative market opinion regarding the use of genetically modified seeds, which practically made this production route unfeasible. Furthermore, it is notable that this type of agricultural approach continues to require chemical products, although in much less quantity. However, organizations devoted to organic agriculture seem to have won the battle in market opinion making, currently occupying a prominent place in relation to biotech agriculture.

The main objective of organic agriculture is to produce healthy and sustainable food using exclusively biological and ecological processes. However, the main problem with organic agriculture is that it is incapable of producing at the same pace as non-organic agriculture, and it is still necessary to produce food for 8 billion people worldwide, with increasing less land available for agriculture. The reduction in area for agricultural activities is essentially due to population growth and the loss of some land due to lack of adequate irrigation (drought), even considering the efforts of engineering to conduct parts of watercourses to areas increasingly difficult to access to that water. Due to the scarcity of organic agriculture products, their price has been almost constantly increasing, which makes this market more prone to be feasible only in richer countries. On the other hand, poorer countries see the supply of organic products as a good source of revenue, due to the price paid by rich countries, turning over of these products and selling them to countries willing to pay a higher price for organic agricultural products, allowing its market to be preferably supplied with non-organic products, more accessible to populations with lower incomes. In turn, more demanding and well-informed consumers are willing to pay more for products knowing that they were based on a strategy of using organic materials as fertilizers in their development, to the detriment of other products that can be harmful to their health, in addition to reducing gaseous emissions, due to the ban on the use of mineral nitrogen as a component of fertilizers, in accordance with the regulation of organic products established by the European Union. Consumers' willingness to pay more for organic-based products motivates farmers to follow this way , using the strategies required for these products to be effectively organic-based products. However, it creates pressure on the supply of organic matter needed for their plantations, which raises the price of organic matter and, consequently, of the final product. Currently, the percentage of the population that is correctly informed and has enough income to have access to food resulting from organic agriculture is still quite low in world terms, which makes this market still a niche. A more accentuated progressive growth is to be expected in the coming years, which will create an accentuated pressure on the producers of this type of goods, considered healthier than those non-organic products. According to the Organic Food and Beverages Market Report, 2022–2030 [17], the market for organic products grew during the COVID-19 pandemic, as people had more time to reflect on their eating habits. Hence, in 2020 there was a significant increase in the demand for these products, which allowed, in global terms, the market for these organic products to reach USD 188.35 billion in 2021. The same report estimated that there will be an increase of around 13% per year in demand for this type of food and drink. However, it is still interesting to speculate whether there would be enough farming capacity to feed 8 billion people worldwide if they all opted for products obtained through organic farming. In 2013, Husaini and Tuteja [18] speculated that organic agriculture would only be able to feed around 4 billion people globally, i.e., only half of the world's population. Given these studies, it will be easy to see that organic agriculture can indeed be healthier, but it cannot meet the needs of food in terms of the world's population. Based on this framework, Husaini and Sohail [19] argued that the only way to escape agrochemicals is to combine older agricultural techniques with biotechnology applied to genetically modified crops. Even so, the challenges to food production for the entire world population are difficult to overcome, so automation and Artificial Intelligence strategies will have to be adopted to increase productivity in this sector of activity. Automation and Artificial Intelligence applied to organic and biotech agriculture can also fill the labor shortage, which is expected to become scarcer when older farmers leave, and new generations become interested in other sectors, less labor-intensive activities. In fact, electronics and new technologies could play an extremely important role in the production and control of agriculture in the near future.

According to data provided by Husaini and Sohail [19], about one billion small holders, mainly in Africa and Asia, are responsible for feeding around

5 billion people, whereas approximately 15 million medium to large-scale farms are responsible for feeding about 1 billion people living essentially in cities. In between, there is a group of 50 million farmworkers on small to medium-sized farms that also feed around 1 billion people. The same authors estimated that if the robotics-based farming system is adopted globally in agriculture, it will reduce agricultural labor to just 0.2 billion workers approximately, who could be solely responsible for feeding around 10 billion people by 2050.

Many of the new technologies applied to other types of activities are perfectly compatible with cultivation activities. In fact, a large part of the activities can be carried out by remotely controlled automatic equipment, sensing can help monitor needs in terms of irrigation, and this can be more directed to the places where it is effectively needed, avoiding the waste of such an essential resource as water. The sensors will be able to feed the information to systems controlled by Artificial Intelligence that control the supply of water and nutrients to the crops, providing them with what is needed, when it is needed, and in the right amount. Monitoring will also allow the analysis of the evolution of products in real time, allowing greater control over subsequent activities, such as harvesting, which can also be largely automated, given the sophistication of existing equipment, and which can also be controlled remotely, even by cloud computing [20].

However, the automation of operations in the agricultural sector also faces many challenges, some not very simple to overcome. In fact, while in a factory building the ground is normally flat and is previously prepared for the installation of equipment, agricultural land can be perfectly level, but this is not common in all cases where automation could be necessary. Different levels and irregular geometries are common in many more rugged regions, which can make the use of certain automatic techniques unfeasible which, in other cases, could be easily applied to agricultural activities. In addition, dust or mud are other problems. In fact, both problems can occur in the same place, at different times, leading to different challenges. The irregularity of the ground with the presence of mud, as well as the difficulty of movement, may affect the performance of the work that was intended to be automatic, causing unexpected difficulties. Dust, in turn, may seem like an easier problem to overcome, as it does not cause unevenness in the ground or in the movement of equipment, but greatly conditions the ventilation systems that would be very necessary in electrical and electronic equipment. This dust, housed in inappropriate locations on the devices, could lead to serious malfunctions and very frequent cleaning and maintenance needs, which implies a lower autonomy and availability of these equipments, removing some of the competitive advantage that they normally have. Additionally, the conditions of humidity and sun exposure during the work period are also quite challenging, requiring that the equipment be prepared for both situations, in different regions of the globe. Given that the agricultural sector still does not have a perfectly adequate sensitivity to new technologies, as well as to civil construction and others, the systems will have to present the necessary robustness to overcome all these problems without making it difficult for those who operate it to conduct the necessary operations that should be performed. Harvesting also represents a big challenge, as the size, positioning and distribution of the products that need to be collected differ greatly. However, for this problem, there are technical solutions that have already been greatly improved, through sensing and artificial vision that could help overcome this problem. Alternatively, the help of an operator can improve the process and reduce costs, costs' reduction being much appreciated by this sector.

The use of Artificial Intelligence is already an extremely important tool, but still very little disseminated due to the technical inability of many farmers to realize the real benefits they can extract from this technology. In fact, most farming operations seem routine, but contain very high variability. The knowledge of data regarding the soil, plantations and harvests, and the improvement of algorithms that, based on the previously obtained results, allow to outline strategies to increase future productions, to know how to react depending on atmospheric conditions, and correctly program the amount of water for irrigation, analyze development time depending on weather conditions, determine the different product development stages, analyze the influence of any products added to each sowing in terms of pesticides, herbicides and fertilizers, crop diseases identification, among many other factors, can be of great interest for farmers. Adequate knowledge of the sowing carried out, soil conditions and corresponding irrigation, and the development of products, are essential in a knowledge society. However, there are sectors more sensitive to these new technologies than others, because of the intrinsic training of people linked to each sector, as well as the training and information provided by the sectoral associations themselves. In certain countries, mainly in low or medium-income regions, agriculture is still an activity that occurs between several generations of the same family, and there is not exactly a company culture, but only a family business. This somewhat hinders the implementation of new technologies, as there are not always specific channels for disseminating technological developments, and how these can minimize the effort and optimize the results of agricultural activity.

Despite these difficulties, the agricultural sector can take extreme benefits from these technologies, that become necessary to disseminate them through associations representing farmers, demonstrating the real benefits they can derive from their implementation. The environmental issue, as in other sectors, is not always seen with the need that ongoing climate change requires. Thus, the much more reliable control that could be carried out through automation, Artificial Intelligence, and other technologies in vogue, which gravitate around the use of real-time information and previous data as a basis for making future decisions, is not always understood by farmers. In fact, from the automatic management of inventories, to controlling the use of pesticides, irrigation needs, weed control, the effect of weather on crop development, control of pollution generated in the activity, water management, and general management of all of its activities can be perfectly automated, feeding algorithms that process data in almost real time and generate performance indicators that allow the farmer to have an integrated vision of the development of his business in its various aspects, also allowing to obtain a greater productivity of the soils and greater competitiveness of products in the market. If the automation of processes implies investments that can be considered large, depending on the type of products cultivated and their diversity, this investment will have a faster return, the larger the area covered by agricultural activity by each farmer, and the smaller the diversity of cultures carried out.

Artificial Intelligence can play a fundamental role in agriculture, as it does not rely on a generalized solution, allowing a much deeper analysis of existing data, provides solutions that consider different scenarios based on the data previously acquired, depending on the learning carried out earlier by the system, rearranging the algorithms to obtain the best solution for each situation. However, as mentioned by Jha et al. [21], computers only really began to be used more intensively in agriculture from 1983 onwards. Trying to follow a chronological order of the evolution of systems applied to crop management, a summary of several studies conducted and presented to apply electronics and computing to the improvement of agricultural production is presented below, although in most cases automation does not directly result in benefits for the environment, but only in improving the profitability of crops, which also translates into benefits for global sustainability, although not focusing only on environmental sustainability.

The pioneer of the Artificial Intelligence use in agriculture was claimed by Lemmon [22], who used a computer to create a model based on an Expert System, called Gossym, which was able to predicting the evolution of cotton crop growth in 1986. The forecast system was essentially based on data from three variables: nitrogen content in the field, the periodicity of irrigation and cotton growth level. In 1987, Roach et al. [23] described the development of a system aimed at helping farmers to dose the insecticide to be applied to fruit trees, more specifically apple trees, to prevent damage caused by weather and insect pests. The model helped farmers choose the most suitable insecticide, as well as the best timeframe. Essentially, instead of following the usual cycle of apple infection in a later phase, the evolution of the disease in the apple was followed. The implementation of the model was successful, considering the opinion of the experts who carried out the first tests of the model.

Still focusing on predicting the growth of cotton fields, Stone and Toman [24] developed a new model in 1989 based on the UNIX operating system installed on a Pyramid 90x computer. The model was essentially based on data from databases, considering two main variables: data related to the field and the cotton crop. Thus, the system was presented as simpler to use by farmers. The system was installed in Texas and based on its predictions, allowed farmers to make the decisions that would allow them to make their crops more profitable. Later, the system would then be prepared for installation on IBM microcomputers and distributed for use by interested parties. In the same year, another model was also presented by Batchelor et al. [25], which essentially intended to measure the effects of insects on the soybean crop, as well as the effect of the amount of insecticide applied on the volume of soybean harvested. The model was also based on previously obtained data, in order to formulate the model based on knowledge. In order to obtain results as reliable as possible, the model was structured in two approaches: in the first one, the results previously obtained by specialists are emphasized, creating their own forecasts, and in the second phase, it filters the conclusions of the specialists, leaving it with its own forecast. It is in this second phase that the most suitable insecticide is found, as well as the flow rate to be applied to the crop. However, the model uncovers an important factor: it cannot predict the damage caused by the insects in crops. In fact, the model should be able to foresee the magnitude of the damage induced by the insect, considering the type of insect, the type of insecticide and the amount administered, studying the economic impact that these factors would cause on the crop. The establishment of specific recommendations for the soybean crop is based on the calculation of the damage rate and the cost to treat the plant, resulting in a given yield.

As early as 1997, Robinson and Mort [26] also used ANN (Artificial Neural Network) to predict and map the ice formation in crop fields on the island of Sicily, based on a broad set of variables and an extensive historical period of values for parameters such as temperature, precipitation, humidity, sun exposure/cloud cover, and wind direction. The existence of these data between 1980 and 1983 allowed training of the ANN algorithm, making it possible to establish strategies with much greater rigor, aiming to overcome this problem. The proper combination of a reliable set of data makes the forecast much more reliable than any other forecast based only on a very restricted amount of data. Studies based on genetic algorithms of artificial intelligence were described by Maier and Dandy [27] who used ANN systems to estimate the variability of available water resources. On the other hand, in 2001 Gliever and Slaughter [28] used ANN to distinguish weeds from crops. Expert systems have also been used in agriculture, but they have some shortcomings when used to predict events. However, if they are complemented with ANN's use, it turns out that there is a strong complement in terms of action, resulting in much more reliable predictions. The same was described by Song and He [29], who used ANN and Expert Systems together to predict in crop nutrition level. The search for a distinction between weeds and crops went even further by Aitkenhead et al. [30], which combined two very useful techniques for this task: artificial vision and artificial intelligence. In this case, the algorithms did not need to be trained with access to previous data, as the Artificial Intelligence worked based only on the formats of crops and weeds, making the distinction. Tests carried out on the system showed that the accuracy obtained in the identification was greater than 75%. In 2008, Singh and Prajnesshu [31] developed an Artificial Intelligence-based model for the efficiency of corn crops. For this purpose, a multi-layer feedforward ANN was used and, to feed this network, learning algorithms were used, such as GDA (Gradient Descent Algorithms) and CGDA (Conjugated Gradient Descent Algorithm), using the MATLAB software.

From a more ecological perspective, Hinnell et al. [32] developed models based on artificial neural networks for drip irrigation of crops. For this dripping to be carried out in the correct proportion, it is necessary to have adequate knowledge of the moisture content already present in the subsoil. The model made it possible to predict the quantity and spatial distribution of water in the subsoil, and the irrigation model was then designed based on this prediction. Based on this prediction, the model could easily make the necessary decisions to turn the dripping on or off, controlling the dripping intensity and the time in which this dripping should occur, to keep the crops irrigated with the amount of water understood as more convenient. Note that in this process, decision-making is extremely fast, leading to much more reliable results. In addition, and in this particular case, only the necessary water is used, depending on the forecast made, allowing an adequate saving of natural resources. In a similar study in 2012, Arif et al. [33] developed a set of two ANN models to estimate soil moisture in rice paddies with the least amount of meteorological data possible. The model was validated by analyzing the difference found between observed and estimated soil moisture values. For developing this first model, only the data referring to the minimum, average and maximum air temperatures were taken as parameters. To develop the second model, more data were considered, such as solar radiation, precipitation, and air temperature. After validation studies, it was possible to verify that both models provided very accurate and reliable estimates of soil moisture in rice paddies using less meteorological data, less labor, and less time expenditure. In 2012, Keshtgari and Deljoo [34] reported the use of the Wireless Sensor Network (WSN) to bring the needs of crops closer to the treatment given to them by farmers. Sensorization allows you to have a real-time notion of the needs of the crops, allowing for quick and effective decision-making. This allows for drastic water savings, as irrigation is not carried out on a temporal basis or based on meteorological forecasts, but with the real state of the soil and its need for water. This technology allowed a significant increase not only in the sustainability of the crops that adopted these systems, but also a strong increase in the quality of the cultivated products. Simultaneously, considerable cost savings can also be achieved. In addition, this type of system is easy to implement and its maintenance is also relatively simple, facilitating its implementation by farmers with less technological skills. It is necessary to carry out a careful selection of the type of sensors most suitable for each farm and crop, but this will probably be the most complex task, since the sensors will collect the necessary information and provide it to a system capable of gathering it, and make the necessary decisions. The flexibility of the system allows its installation to be carried out in a short period of time and it can adapt to any type of cultivation.

With the popularization of smartphones, applications began to be developed that allow estimation yields based on ANN, as described by Ravichandran and Koteshwari [35] in 2016, in which the forecast accuracy depended essentially on the number of hidden layers. Initially, different algorithms were used to train the model, such as Rprop, Delta-bar-denta or de Silva e Almeida, in order to find which configuration was more efficient in forecasting. Given the lack of knowledge for defining the most appropriate number of hidden layers, the trial-error method was used, which allowed to verify that the greater the number of hidden layers used, the more accurate and reliable the prediction method became. Aiming to make the system as easy as possible for farmers, the model was developed on an APK platform, with the source code written in Eclipse with Java codes. In turn, the algorithm was developed using Matlab and ANN toolbox. Subsequently, the software was adapted for the Android platform, allowing its easy use on most smartphones. The model was able to indicate when farming the crop was to be carried out, also indicating which fertilizer was most suitable for that crop [35].

Also considering environmental sustainability and how automation and electronics can help to improve the environment, by saving resources, it is important to understand how the evapotranspiration process works, which is essential to achieve the required stability of the hydrological cycle, sustainable irrigation method and water management. So that a model can be developed in order to optimize these resources, it is necessary to have access to data such as the elevation of the land, the minimum, average and maximum daily temperature, wind speed, relative humidity, the average number of hours of sun and light each day, the latitude, and a coefficient that depends on the condition of the terrain and crop. However, evapotranspiration can be determined by more than 20 different methods, which depend on different parameters. To establish a prediction model for evapotranspiration, Nema et al. [36] collected climate data from a given Indian region, with a view to integrate these data and estimating evapotranspiration using two methods: the Penman-Monteith method and the Levenberg-Marquardt Backpropagation method. Contrary to what was observed by Ravichandran and Koteshwari [35], in this case, an increase in the number of hidden layers led to dispersion and lack of rigor in the estimates. It was observed that of the six training algorithms of the ANN model, the function training with 75% of the data feed was the one that provided the most reliable results and presented the best number of neurons. The ANN model was designed and developed based on Matlab software. Six different algorithms were combined and evaluated. As evapotranspiration is of vital importance in irrigation and water management, this investigation provided a much more reliable estimate of water requirements through a correct prediction of the hydrological cycle and evapotranspiration. In 2012, and using much less sophisticated techniques, Ingale and Kasat [37] used a simple 89c52 microcontroller IC to control a smart irrigation system. The prototype supplies water only when the humidity drops below a certain standard value, thereby optimizing water consumption.

Recently, another extremely useful technology for agriculture has been developed, which is based on thermal imaging. This technology is non-invasive and contactless, allowing the collection of extremely valuable information for the farmer, through the collection and analysis of the surface temperature of the crop. This technology is a powerful aid for farmers in the irrigation process, defining the amount of irrigation depending on the temperature in each location, through the temperature mapping carried out, also defining the best timing to do so. Even if there are asymmetries in water needs, the system can perfectly manage this, irrigating more areas with higher temperatures, and less those with lower temperatures, making it possible to define the temperature from which irrigation is necessary, or program the system so that the decision is taken automatically. This technology is extremely sustainable, as it only consumes water strictly according to the needs defined by the system [38]. However, this technology had already been considered before in order to prepare operations such as pre-harvest operations, field nursery, irrigation scheduling, yield forecasting, greenhouse gases, termite attack, farm machinery [39].

Yong et al. [40] introduced a new concept called Embedded Intelligence, which, translated in agricultural terms, includes smart farming, smart crop management, smart irrigation, and smart greenhouses. Currently, given the constant development of technology, GPS systems have also been used to allow the remote control of robotic systems capable of carrying out certain agricultural tasks, such as harvesting, localized analysis of humidity, spraying, surveillance, bird scaring, among them , many others. But, even using much more economical technologies, based on small microprocessors such as the *Raspberry pi* or the *Arduino*, in parallel with some sensing, it is possible to have a much more rigorous notion of what is happening globally in the entire cultivation, and act locally where necessary. Thus, with very economic means, it is possible to take advantage of technology to save increasingly scarce resources such as water, since the consumption of water in agriculture represents a huge share of the global consumption of drinking water on the planet. It is foreseeable that this situation will evolve further, given the increasingly easy access to more developed technology, namely, drones. In fact, drones, associated with artificial vision or thermal imaging technology, can give a new impetus to agriculture towards becoming more environmentally sustainable. The automation of ways in the same farm, associated with thermal imaging or artificial vision, and coupled to computer-controlled irrigation systems, will make it possible to identify the water requirement in each zone, and proceed with the irrigation of each zone in the exact amount needed, saving resources. Concerning disease prevention in crops, real-time analysis carried out by artificial vision systems coupled with drones will be able to create the necessary alerts for localized action, in order to prevent the spread of diseases throughout the crop, and conduct general or localized spraying, depending on the needs. Sowing and harvesting operations can be identified in terms of adequate timing, but they can be carried out automatically, as they are already carried out in most developed countries, and even in developing countries. However, most farmers do not have adequate training to handle these technologies. Thus, there is strong pressure in the market for professionals who are skilled in handling new technologies, since in most cases it is necessary to carefully select the technologies to be implemented, the integration of these technologies into a common platform with relatively easy control, a programming appropriate to each case, and the establishment of management indicators and alerts that allow adequate control of the situation experienced on the ground. Furthermore, the maintenance of these systems also requires expertise in new technologies, so technology alone is very useful, but not enough. The human resources required for the integration and maintenance of agriculture automation systems will be increasingly in demand in the future. In fact, including agriculture, which are arduous tasks to carry out, and with a lower predisposition of the new generations to perform-these tasks, the only way will be automation, which is can attract young people who have a greater appetite for these technologies and see in them an exit for the continuity of the exploration of farms previously explored by their families. Believing that the new generation has greater sensitivity to environmental issues, and that legislation will evolve and restrict activities harmful to the environment, there is also a great opportunity to make the agricultural sector less polluting and more environmentally sustainable.

Regarding the use of automation in agriculture, Tian et al. [41] presented a work that summarizes the research work presented in recent years, mainly taking into account the use of artificial vision, segmenting these works by focusing on each phase of the agricultural process, such as crop growth monitoring, disease control, automatic harvesting, quality testing, automated management of modern farms and the monitoring of farmland information with Unmanned Aerial Vehicle, showing that the studies practically cover all tasks related to the agricultural process, different solutions exist for each one of the challenges presented by the activity. The biggest challenges for the progression of automation in agriculture are essentially the need for people with adequate training and a comprehensive vision for integrating different technologies into a single system capable of analyzing and making decisions based on the previously defined algorithms, and the availability and robustness of the necessary equipment, considering the agricultural environment, exposed to a wide range of climatic conditions, and some situations that are difficult to manage for more advanced technological systems, such as wind, dust and mud.

Robotics applied to agriculture is usually more associated with sowing and harvesting than with other tasks where environmental sustainability needs to be improved. However, through systematic operations, both the sowing and harvesting tasks can be optimized, allowing lower levels of energy consumption and, consequently, less pollution. Studies carried out recently have led to the development for commercial purposes of a series of robots that perfectly adapt to certain agricultural tasks, as can be seen in Fig. 3.1. The work presented by Gil et al. [42] briefly describes the different approaches taken by various researchers in the development of robots designed strictly for research purposes, as well as a summary, but quite interestingly, the description of a series of robots developed with perfectly commercial purposes, and which are either intended to be available on the market to be adopted by farmers, depending on the specificity of each robot and the type of operations and crops for which it is intended.

The supply of robots for agriculture on the market is very diverse, with different dimensions and types, which depend on the functions that these robots need to perform. The drive is also very diverse, from omnidirectional, Skid-steering, 2WD, 4WD, Akermann, Crawler, and even differential three wheels among others. Displacement speeds are normally between 1 and 4 km/hr, and autonomy is between 4 and 20 hr. Most are equipped with GPS, which is the technology that guides the remote control. They are also equipped with numerous sensors of different types, which allow information to be collected on the ground and sent to the control systems, where the software processes the information and generates decisions, which are transformed into actions transmitted to the robot, and this takes place on the ground. Note that most of the robots are driven by electrical energy accumulated in batteries, and some of these robots are already equipped with solar panels capable of greatly assisting in charging the batteries, making the systems much more environmentally sustainable.

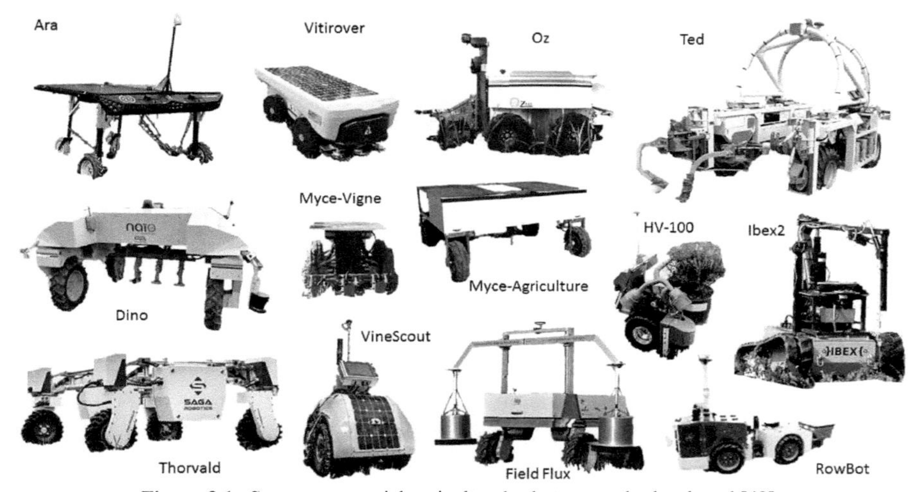

Figure 3.1. Some commercial agricultural robots recently developed [42].

The next step, already intensely discussed in terms of research, is to make these robots autonomous in terms of the decision-making power, presuming they act under the control of well-determined algorithms and with high security conditions, imposed by redundant systems that avoid any type of failure, or immobilize the robot in the case of loss of control. However, although these robots operate in environments that are rarely used by humans, their interactions can be catastrophic in the event of a failure, so there are still many situations to be discussed and possible adjustments to be made before the agricultural tasks carried out autonomously by robots can become a reality.

However, it is perfectly clear that agriculture will evolve immensely in terms of technology, and that autonomous agricultural tasks are no longer fiction, being only dependent on creating the appropriate safety conditions for their possible interaction with humans in their tasks.

It is also clear that automating these tasks can significantly contribute to greater environmental sustainability. It is also evident that a large part of the food resources generated by agriculture are consumed by the agriculture and livestock sector, that is, for raising meat. For the market to move more focusedly on environmental sustainability, there needs to be adequate food education, which favors the consumption of vegetable products to the detriment of animal protein, since the number of crops needed would be lower and progress that would be made, move in a more sustainable way towards greater environmental sustainability [43].

Although agricultural crops constitute the largest area and, therefore, attract greater attention concerning the consumption of natural resources, there are other areas that should be governed by the same principles, as they suffer from the same problems [44]. In fact, large public parks, golf courses and even private gardens, in order to be maintained in the conditions normally required, also need water and energy. If the same precautions previously indicated for crops are applied, these installations could also contribute in a very positive way to increasing environmental sustainability through the automation of processes. In these cases, irrigation and weeding are the biggest problems, but these are common to what was previously analyzed for agricultural crops. Regarding irrigation, and including the image of agricultural crops, irrigation, in addition to water consumption, also entails energy consumption in pumping. Here, too, automation could play a key role in improving environmental sustainability, through the generation and management of electricity based on solar panels, which manage the energy normally consumed in pumping. Since irrigation is normally carried out during hours of exposure to the sun, energy management can be carried out in such a way that there is a direct use of that energy, or that there is an accumulation and management of these energy resources, to use it as efficiently as possible.

3.3 Automation in construction

Buildings are often architectural landmarks that define the silhouette of a region, giving it its own identity. For this reason, many places are very careful in the architectural definition of each building, licensing only those that fit the intended identity. As it is something that has been preserved over time, and because these buildings are something that can bring happiness to those who live in that region, and even attract tourists to visit it, many buildings easily exceed a century of existence, i.e., even for their cost, are preserved over many decades. However, the needs of those who live or work there change over time. The requirements in terms of comfort, technology and environmental sustainability today have completely different requirements from those assumed at the beginning of the 20th century. If the exterior appearance is normally preserved because it is part of the identity of a region, its interior needs to undergo significant updates, accommodating all the evolution felt. In recent decades, the awareness that the non-fulfillment of some requirements in the construction would contribute to a serious gap in the environmental and economic sustainability of buildings, has changed the panorama in this sector.

The 2022 UNEP report on the construction sector is very clear: despite all efforts to increase energy efficiency so far, the sector is not able to meet the goals set for the 2050 situation in 2021, the decarbonization index of buildings is situated at 8.1 points on a scale of 100, when it should be above 17.1 points on 100, to comply with the objectives established by the Paris agreement. The figures show that the decarbonization effort is just over 50% of what is established in the agreement. Additionally, the gap between the current trend and the one needed to comply with the agreement has been widening since 2018. To make the scenario even worse, it seems that no structural and systemic improvement is truly designed and being implemented in the energy sector construction, leaving it vulnerable to external factors such as fluctuating consumer prices, inflation, and temperature changes.

The energy consumption required by poorly insulated buildings can be a large expense over the entire life of a building. Therefore, building design should include a rigorous study of sun exposure conditions to minimize energy consumption during its useful life. Many older buildings have undergone improvement works with a view to essentially improving their energy efficiency, care that has been absorbed by the political class, especially in more developed countries, establishing strict legislation that obliges buildings to comply with standards capable of making them more sustainable. The awareness of the need for decarbonization and, more recently, the ever-increasing cost of energy, has accelerated this process, as this is a sector where it is easy to see that most of those involved, whether in the design phase or in the construction phase, or even as consumers, they know what to do to comply with the legislation and generally seek to provide products as sustainable as possible to the market [45]. According to estimates pointed out by Lin et al. [46] in 2009, the construction sector was responsible for around 30% of global greenhouse gas emissions and was also responsible for consuming around 40% of the global energy market. In addition to the energy consumed in construction, office buildings consume the most energy, essentially due to heating, ventilation, and air conditioning. This has led designers and builders to try to establish as much automation as possible in buildings, creating, on the one hand, the best working conditions, which are reflected in greater productivity, and on the other hand, trying to minimize costs throughout the life-cycle of the building, through energy management with a faster and more efficient reaction, which is also more environmentally friendly. Although the sensitivity for constructing intelligent buildings has increased in recent decades, both in terms of design and construction and renovation of buildings, the number of studies around the subject remains quite modest. When an intelligent building strategy is adopted, the main concerns are centered on making the building as self-sustaining as possible, through renewable generation of part or all the energy consumed by the building, as well as adopting energy saving strategies that can reduce the energy consumption throughout the useful life of the building and application of products that contain recycled materials or that facilitate future recycling. According to a report by the European Commission [47], the implementation of assertive strategies for smart buildings can lead to water savings of around 30% and savings in electricity consumption of around 30%. These strategies can include the installation of sensitive panels on the exterior walls of buildings, and strong sensing inside that provides the necessary data to artificial intelligence algorithms that make decisions depending on the set of data collected at the time. In view of the conditions felt inside the buildings, the facades can assume different configurations, intervening in the process of heating or cooling the envelope, thus minimizing the effort of equipment installed inside, which consumes energy to establish the necessary comfort conditions.

In the European Union, the LIFE program, established in 1992 [48], strove to boost environmental sustainability in the construction of buildings, with very significant results. In fact, buildings can constitute a significant milestone in the development of sustainable practices, which can range from the simple installation of solar panels for one's own consumption or supply to the grid, to the automation of a large part of the building's functions, allowing energy savings, in addition to being designed in such a way as to assist in the segregation of solid waste with a view to greater ease in recycling. Moreover, the unstoppable succession of innovations in the technological aspects has allowed building designers to develop solutions in the most diverse aspects of environmental sustainability, including the reuse of water. Building design can also contribute to increase environmental sustainability through the preferential use of natural or locally produced materials, avoiding pollution generated by logistical processes. Furthermore, some materials that are difficult or impossible to recycle, and that do not represent toxicity for the environment and for the users of the buildings, can be integrated together with mortars, preventing these materials from accumulating in sanitary landfills. In fact, the development and selection of the best materials considering all aspects related to construction, such as architectural aspects, sustainability in environmental terms, comfort, and cost, have been the subject of many studies and the progress of environmental awareness, is well known in this sector, which, through examples of good practices, has stimulated the improvement of techniques and technologies, to become increasingly capable of providing truly sustainable solutions in environmental terms. Architecture schools and sectorial associations have also played a leading role in the evolution registered so far, disseminating advanced practises and explaining the goals behind the ideas.

In order to demonstrate how buildings can be sustainable using existing technology, a project developed under the LIFE initiative, called EDEA, which constituted a comparative demonstration of how energy efficiency can be explored in construction. For this purpose, two houses were built in the same location, with the same geographical orientation and subject to the same weather conditions, one following the most traditional construction principles, while the other incorporated a series of technological innovations to increase environmental sustainability. In this innovative concept, several renewable energy generation technologies have been used. With a view to the comparative study, around 300 sensors were placed in both houses, having adopted a set of 15 different passive strategies and 70 active strategies. It was stated that the passive strategies contributed essentially to the improvement of the construction and design of the building, while the active strategies contributed decisively to the improvement of energy efficiency and reduction of CO_2 emissions [48]. A similar project in environmental terms, but aimed essentially at the use of new technologies for the automation of buildings, was also carried out under the name DOMOTIC, with the main objective of saving energy consumption and minimizing CO_2 emissions. The implementation was carried out in three different buildings, with different years of construction and using different techniques and construction materials, but having one characteristic in common: they were very inefficient in terms of energy, consuming significant amounts of energy. By installing different technological devices, from motion sensors to integrated interior temperature and lighting controllers, it was possible to achieve savings in energy consumption of around 50% for HVAC systems, and around 80% in terms of lighting, when compared to a traditional approach. In the set of the three intervened buildings, it was possible to improve the annual energy efficiency by 63.9% and, through the implementation of renewable energy, it was possible to avoid the emission of 680 tons/year of CO_2, which represents a saving of around 162,000 €/year for all three buildings and a return on investment in just 4 yr.

More recently, the European Union started to encourage the application of automation and robotics in the construction of buildings, focusing essentially on reducing harm to the environment. With this objective in mind, the European Union has allocated funds to boost automation in construction and prefabrication, provided that they significantly improve energy efficiency and use environmentally friendly materials [49]. At the same time, the concern with waste from construction and the reduction in the consumption of natural resources has been a recurring concern of companies whose top management is more sensitive to environmental issues. The demolition of buildings is another matter of enormous concern within the construction sector, as it generates a huge volume of waste. This waste, in most cases, does not follow a perfectly defined principle of separation of materials, later facilitating reuse, reconditioning, recycling or landfilling operations. However, in demolition operations, the use of automation and robotics has not registered any significant evolution, given the specificity of these operations. Only companies with procedures already perfectly outlined carry out a more organized dismantling of some structures, such as the planned removal of metallic structures, when the buildings have characteristics that allow this type of approach. It should be noted that the use of very dissimilar materials, sometimes deeply connected, due to the nature of the weather to which they are subject during their useful life, makes properly organized dismantling extremely difficult. It is worth noting that some masonry structures are almost impossible to recover, so specific reconversion procedures are required.

Pan et al. [50] carried out a study that aimed to create a framework that would allow in understanding the degree of sustainability of a given building still in its design phase, in which the use of automation and robotics was foreseen. The study concluded that the use of automation and robotics needs guidelines that lead this application in the direction that is really intended, that is, in the perspective of reducing the environmental impact. To this end, it also developed an indicator that allows quantifying the contribution of the planned application of automation and robotics, identifying whether this contribution is within the sustainability parameters that are required for agreements on the preservation of the environment to be effectively complied with.

It can thus be said that construction is a sector that can contribute in a very significant way to environmental sustainability, mainly through the use of automation and robotics, but there is still much to be done in this regard, although there is already vigorous regulations in some countries, and the companies also already have some sensitivity to environmental issues, as they are also pressured to do so by the market.

3.4 Automation in industry

When talking about pollution and environmental degradation, one of the images that immediately comes to mind is that of factories with their chimneys belching smoke. Industry involves a significant portion of other sectors around it, all of which contribute to environmental degradation. The same takes place with the energy sector, where a significant share of consumption is also directed towards industrial needs. In fact, industrial activities are responsible for about 90% of the energy consumed worldwide [51], which implies the generation of greenhouse gases. Numerous efforts have been made to minimize the consumption of electricity in industry, given the facts described above, and all the negative burden that this entails on the environment. While in construction there is its own logistics for materials, a large part of the work is carried out locally, the globalization of the industry has also produced extremely negative effects on logistics chains, which are responsible for a significant share of greenhouse gas emissions in moving materials and parts among stakeholders and/or business partners. This means that, even though logistics is considered a service activity, it is also driven largely by industry. However, the industry itself also has its share in producing extremely negative effects on the environment.

Perhaps because it is an extremely competitive sector, in which products are constantly under the scrutiny of consumers, it could be the sector where the most efforts have been made to minimize environmental impacts, but probably the one where more efforts are still needed in the sense of reaching values compatible with the political commitments assumed by many countries for 2050. Given the diversity of subsectors within the industrial sector, there is a much more accentuated development, extremely subdivided, but which is reflected in a large number of studies in the most diverse areas. In most cases, environmental sustainability develops in parallel with economic and social sustainability, as the aim is to reduce the energy consumed, production time, and the risk of accidents for workers, or to minimize occupational illnesses. A good part of the studies performed to increase economic sustainability can also contribute to environmental sustainability, as saving natural and energy resources is a permanent concern for controlling costs and increasing the competitiveness of products, and this significantly directly influences ~~to~~ environmental sustainability. Sometimes, some studies do not refer to the importance that a certain development has for the environment, referring only to the component of economic sustainability, as this was the focus of the development of the work. We all hope that this situation will one day be reversed, with the work being carried out with a focus on the environmental component and, at the same time, also benefiting economic sustainability.

Automation and robotics have contributed very effectively to a very positive trend in the environmental sustainability of the industrial sector. Considering the information provided by the International Federation of Robotics, in 2021 more than 517,000 new robots were installed in the industry, surpassing for the first time the level of 500,000 robots installed for 1 yr [52], about three times more than the robots installed worldwide in 2013. These robots are usually installed taking into account different factors, such as guaranteeing the performance of very repetitive and tiring operations for the human being, replacing the human being in operations with high safety risks (need for working close to high temperatures like in foundries or hot-stamping processes, for example), processes where it would require a lot of energy on the part of the human being, processes where it is needed to increase repeatability and guarantee greater levels of confidence in quality, increase productivity, among many other factors. However, to these should be added with particular emphasis on the possibility of robots being applied to specifically contribute in a positive way to environmental sustainability, saving resources and greenhouse gas emissions into the environment. In fact, it was mentioned by Fysikopoulos et al. [53] that the energy consumed in the production of a car represents about 9 to 12% of its global manufacturing cost. If one considers this percentage and it is possible to reduce the energy spent to manufacture a car by 20%, the overall production costs of a car could be reduced by about 2 to 2.4% just by reducing the energy consumed in the process of production.

Lofving et al. [54] mentioned that in 2018 in Sweden, the degree of automation in production in companies that produced essentially small batches of products was less than 1%. Perhaps it would appear surprising, but this probably does occur in any other developed country. Initially, automation was only of the fixed type, so the installations were designed for a given product, with zero flexibility to produce different versions. Indeed, in terms of industrial production, fixed automation was formerly seen as a valid solution only when the series to be produced were so large that the installation of an automatic production line for just a single product was enough to ensure the return on the high investment needed. Fixed automation is a very interesting solution when the prospect of producing a single product is sufficient to ensure a secure return on investment. However, the need for companies to customize products significantly increased the range of products to be produced by each company within the same family, but with manufacturing specificities that prevented fixed automation from being adopted. Given the high costs previously referred for such automation, it is not a solution for small batch production, so the evolution from fixed automation to flexible/reconfigurable automation is an excellent alternative for all companies that should evolve towards the automation of their production lines [55-57], but they could not do it through fixed automation, due to the diversity of products they produce. With the possibility of programming certain devices widely used in industrial automation, it has become possible to make systems significantly more flexible, with automation starting to be considered a precious auxiliary for manufacturing in small batches, according to the characteristics intended by the customer, due to the possibility of easy reconfiguration of processes, significantly streamlining the change of production from one product to another, i.e., significantly reducing setup times and manual operations around the reconfiguration of the necessary tools, etc. [58-63]. Currently, conventional automation, through the possibility of programming, aided by code reading systems and even artificial vision, has enabled a very significant advance in the last two decades in most industrial operations. The use of flexible/reconfigurable automation is, by itself, an environmentally sustainable solution. In fact, the flexibility provided by these systems allows the same equipment to produce a series of different products, as long as they are minimally identical, making unnecessary the use of different equipment, one for each type of product. This fact makes it possible to optimize the use of the material used in the production of this reconfigurable automatic equipment, reducing the consumption of materials, therefore, of resources. On the other hand, it makes the return on investment easier, allowing process automation to proliferate more quickly. If this automation allows saving resources, namely in terms of savings in terms of energy consumed, less waste of materials, higher quality indices and, consequently, less scrap, also reconfigurable/programmable automation constitutes an extraordinary way to increase environmental sustainability. The main focus remains the same: increase productivity. However, some studies question how increasing environmental sustainability can be found.

Conventional automation, both fixed and reconfigurable, also has the advantage of being made up of several components that are multifaceted, allowing its application in different contexts, as long as its function remains the same. Given that the life of a product in production is not very long, varying depending on the competitors' evolution and market requirements, the production equipment has roughly the same duration (it may be longer if the manufacturer needs to manufacture the same product for later replacement). This implies that production lines may no longer be needed at the end of that period. However, most automation components are not obsolete at the end of this period and can be reused in future projects. Some studies have revealed that this is a current practice, given that companies should monetize their assets [59,62,63]. Although the focus of this reuse is essentially economic, it is very significant in environmental terms, as it avoids consuming new resources on the one hand, and on the other hand, it avoids encouraging the recycling of a product that can still be used. Thus, even if not explicitly, automation can also contribute to a better environment in this way. For this practice to be even more consistent, increasing the useful life of automation devices should be encouraged. However, this reuse is not always possible due to the strong evolution of technology, which makes certain devices obsolete or not very interesting in terms of performance.

Modularity in automatic systems is something that also significantly facilitates the reuse of complete systems or modules. When designing, systems can be divided into cells, promoting the division of operations into workstations [64]. In this way, when there are modifications to the product and this brings more severe implications for the manufacturing process (other than just changing tools), one or more modules can be removed where the product modifications have the greatest impact, keeping the remaining modules in their initial setting. Associated with what was mentioned earlier, these practices can also make automation more environmentally sustainable, since, in addition to contributing to a better environment in terms of energy savings and consumption of natural resources in the manufacture of the product, the system itself that allows the automation of the process is also more sustainable, allowing it to be reused more quickly.

According to published data [65], about 38% of robots are used in packaging and palletizing operations. Even in this aspect, significant gains can be achieved for the environment, since the electrical energy consumed by the robots can be produced by renewable means and, even if this is not the case, the pollution generated in the electrical energy consumed will be used more efficiently and in a smaller amount than if these tasks depended on operators who carried out the pallets with forklifts or other equally polluting means of movement. In actual production operations, welding is the manufacturing process where robotics is most widespread, with around 29% of robots installed worldwide being allocated to this type of manufacturing process [65]. The reproducibility of the robots ensures greater quality in the creation of weld beads and greater efficiency in the energy consumed. It is therefore understandable that robots are widely used in welding processes because the operations are routine, and the involving processes of positioning and removing parts from the jigs can also be automated [66]. In this way, routine work is transferred to the equipment and the necessary skills are increased, as the programming of robots requires specialized personnel. In terms of environmental sustainability, the lower number of defects produced by the robots saves material resources and the non-need for repair and repeat operations due to operator carelessness, leading to lower energy consumption. The simple optimization of trajectories in welding, or even in laser or waterjet cutting processes, can significantly contribute to energy savings, which contributes positively to the environment. However, environmental concerns are not probably the main motivations for applying robotics in welding, as the productivity, repeatability and quality ensured by robotics are far superior to what any human being can do at their workplace, where they are subject to fatigue and various types of stress and unforeseen events. Robotics is used in many other processes, but after welding, the operations that most use robots are painting, gluing, and spraying, sectors with around 10% of the total number of robots used worldwide. All these operations deal with chemicals, which are normally harmful to health and the environment. The use of robots clearly increases social sustainability, as it removes workers from functions where they would be clearly exposed to products that could cause health problems. On the other hand, the systematization of operations contributes to an increase in quality and productivity, which translates into an improvement in economic sustainability. Finally, and undoubtedly most importantly, the effectiveness of the paths defined through effective and rational programming will avoid wasting ink, thus saving material resources, and avoiding higher levels of pollution. The increase in quality minimizes waste of materials that are difficult or even impossible to recycle. Therefore, robotics, associated with automation, represents an important step in increasing environmental sustainability, saving the use of resources, a higher level of pollution and greater energy consumption.

Studies have also been carried out on the improvement and control of the geometry of the weld beads when performed with the aid of robotics [67,68].

Apparently, this may seem like a problem of little relevance to the environment, but here automation can also become a fundamental tool for minimizing the effects of electric arc-welding processes on the environment. In fact, by adjusting the weld beads to needs, one is contributing to the elimination of the need for future deburring operations, in addition to the presumed material savings in making the initially deposited bead. By avoiding grinding, the use of materials harmful to the environment is avoided, as abrasives are made up of products normally harmful to health and the environment. By not having to use abrasives, one is also saving the use of resources, which, in this case, have direct implications for the environment.

Although robots need energy to move, and the energy savings they generate can be controversial, there are many factors surrounding robots that allows one to understand that, despite the energy they consume, they allow other savings that largely compensate its use. In fact, a work environment that is largely robotized needs a much less intense level of lighting than that required by the work carried out by humans. In addition, the heating or cooling normally required by the human workforce is much higher than that required by robots, which also have requirements at this level, but are less demanding. As far as the operation of the robots itself is concerned, the use of lower speeds and accelerations does not linearly lead to a reduction in energy consumption, quite the contrary. Therefore, the speed of the robots should be selected and optimized according to economic sustainability, increasing the speed without any concern for consumption, because the higher the speed, the lower the consumption. Moreover, several studies have been conducted to optimize the energy consumed by robots [69]. Wang et al. [70] carried out a study based on 38 different countries to understand how the consumption of electricity was influenced by the adoption of robotics in industrial processes. After collecting all the data and carrying out robustness tests regarding the results obtained, the authors confirmed that the adoption of robotics contributes significantly to the reduction of electricity consumption, thus being an environmentally sustainable way. They also concluded that robots are applied mainly in industrial sectors where intensive labor prevails. As previously mentioned, the comfort that should be provided to operators is a factor that significantly affects the consumption of electrical energy. Thus, it is easily concluded that the adoption of robots to perform repetitive tasks previously carried out by operators reduces the energy consumed, since the energy consumption of the robots is much lower than the energy needed to efficiently conduct the conventional processes, as well as to provide the comfort in terms of lighting and heating/cooling needed by operators carrying out tasks using conventional processes. The efficiency of robots can be greatly improved through proper production planning. So that the energy consumption of the robots and all the peripheral automation can be optimized, it is necessary that the interaction between the robots should be properly balanced, balancing the workload between different robotic cells. In this way, the effective working time of the entire set of robots is optimized, allowing neither time to be wasted nor energy consumed unnecessarily, which translates into unnecessary consumption of resources, a situation that is intended to be avoided as it is harmful to the environment. The balancing of production lines, both in conventional and robotized processes, is an essential procedure to guarantee an effective workload balance and, therefore, greater energy efficiency [71].

Commercial software, commonly available for programming robots is primarily, focused on optimizing the production time. However, some software already allows priority to strategies aimed at saving energy consumed in the process. However, it is still possible to optimize the strategies provided by the commercial software, optimizing the trajectories also considering the other devices added to the robots. When a robot is being programmed, there is a significant difference between what is called a "path" and a "trajectory". Effectively, when the term "path" is used, it is just defining a given geometric line to be followed by the robot. When referring to "trajectory" programming, the path to be taken by the robot is established as a function of time, i.e., for a given path geometry, a specific velocity/acceleration profile is established over time [72]. Trajectory strategies aimed at reducing energy consumption can essentially be based on two principles: (a) eco-design, in which energy savings are achieved based on the application of more energy-efficient hardware, or through the remodeling of the productive layout; (b) eco-efficient programming, in which the layout is adapted without any significant hardware change, essentially considers the robot's programming and adjustments in the production scheduling. With a view mainly to saving energy in operations carried out by robots, Merlo et al. [72] developed a new programming style for 6-axis robots, optimizing their trajectories, regardless of which process the robot was operating in. To this end, considered all the losses in the various systems that make up a robotic cell, such as electrical motors, actuator drive systems, and controller cabinet. The Dynamic Time Scaling strategy was used to design the algorithm, with energy achieved savings between 2 and 13% compared to other previously developed and published programming methodologies.

More recently, collaborative robots have emerged, which allow greater integration and interaction between automated work and manual work. In fact, the abrupt way in which industrial robots move and any gap in programming could lead to sudden movements of the robot, which could endanger the operator, considering he was within the robot's working volume. To avoid these risks, industrial robots are usually surrounded by their own safety barriers, preventing any operator distraction from causing a serious accident at work, as the robots perform movements just following the programming that leads them, without checking for any obstacles, human or not, stands in its way. Collaborative robots have brought some advantages in this regard, as they are equipped with corresponding sensing that prevents the robot from colliding with people or objects that get in its way. The ability to perform some jobs is more limited than in traditional industrial robots, but, on the other hand, it expedites other perspectives in interaction with humans. Thus, it can be stated that the field of application can be significantly different, and each type of robot has advantages and disadvantages. However, it allows new usage perspectives, which would be difficult to achieve with traditional industrial robots.

The use of simulation in manufacturing processes such as welding, among others, makes it possible to predict and avoid certain production errors that, if not avoided in this way, will lead to the production of non-functional prototypes. All material produced that is not functional translates into a problem for the environment, as it consumes resources and is of no use, being necessary to convert it, recycle it or send it to landfill, with the due consequences for the environment. It can therefore be said that the development of simulation software is a big advantage for the environment, as it can significantly bring prototypes closer to final solutions, without significant waste of resources. Simulation, which can also be considered an automation tool, can help optimize the geometry of components and assemblies, which can lead to significant savings in resources. The development of software dedicated to simulation has seen an extremely positive increase, as the advantages offered by using this software are extremely important for economic and environmental sustainability. Saving material immediately translates into saving money (economic sustainability), but also saving resources (environmental sustainability). In fact, some over-dimensioning carried out to safeguard trouble-free operation during the useful life of a given product, can now be consciously optimized, without jeopardizing the useful life of the products, but allowing to save resources, which helps safeguard the environment. The cost of simulation software has dropped significantly over the past two decades, and it has even become integrated into many 3D CAD (Computer Aided Design) packages. The integrated use of CAD and CAE (Computer Aided Engineering) brings significant advantages in product development, allowing the optimization of geometry and consumption of materials that specifically meet the conditions of use. On the other hand, it appears that in capital-intensive companies, the reduction in energy consumption with the implementation of robots may not be as significant, or even have a negative impact. However, when the phenomenon is correctly dissected, it turns out that this higher energy consumption is due to all the technology that gravitates around the processes where the robots are implemented. In fact, this type of industry resorts to electrical energy for many other systems that become necessary for the industrial workflow that these companies have, so that the increase in energy consumption does not come from the robots themselves, but from complementary systems that would have been equally necessary if the processes had not been robotized.

In fact, automation and robotics can contribute to a greener industrial sector through different solutions, as mentioned above. From simulation in the design phase (CAE), which can be worked together with 3D CAD during the initial phase of product development, passing through software for simulation and optimization of manufacturing processes, such as CAM (Computer Aided Manufacturing), whose strategies could specifically lead processes towards more environmentally friendly solutions, and ending with the definition of manufacturing processes where automation and robotics provide significant advantages for the environment in relation to conventional manufacturing processes, the industry can assume its share of responsibility in reducing environmental degradation, as it has the right tools for this purpose: the use of automation and robotics on an ever-increasing scale. Due to the importance of the industrial sector, the next chapter will look in more detail at how industrial processes can be made greener.

3.5 Automation in services

In recent decades, the tertiary sector has gained a particular notoriety. Technological development, especially in more developed countries with better wages, has promoted a strong increase in services, a situation that would certainly have been unimaginable five decades ago. The development and trivialization of the use of computers, the creation of the internet and the democratization of its use on various devices, and then the appearance of mobile phones and smartphones, triggered tremendous needs in terms of software and other services. If the computer or the smartphone is our window to the world, it is all due to the development of increasingly friendly software, more intuitive to use, which has allowed an easier and faster adaptation of older generations with less sensitivity to new technologies.

Concerning the tertiary sector, it can be stated that a service is effectively green when it is able to designing or reformatting a company's operations in accordance with new environmental concepts, or even with environmental legal restrictions with a view to adopting sustainable operational paradigms. In fact, when the service sector focuses on making its operations truly sustainable, it has the necessary potential to reshape the product's life cycle, reformulating operating standards and minimizing the negative effects that a product considered at the end of its useful life could have in environmental terms. Thus, sustainable practices can be stimulated by extending the useful life of the product due to new functions developed by the service sector, reinventing ways for the product to remain useful beyond the estimated initial period of its useful life, namely, through software updates, among others [73,74]. To make the service sector's strategy more focused on the environmental sustainability, more digital systems should be adopted able to promote the enhancement of the offer of products coupled with services based on new intelligent and disruptive Information Technology solutions in relation to technologies that are less environmentally friendly. Software is just one of the many areas where services now rule. Hence, it is the first topic to be addressed in this chapter.

One of the questions that can immediately be asked is the following: can software help improving environmental sustainability? The answer is obviously positive. Software is largely present in almost all the electronic devices we use. From a simple toothbrush to our car, the devices' actions are controlled by us, but using an interface and algorithms that are based on increasingly user-friendly software. However, the software is not limited to personal applications in our day-to-day life, but also in industrial situations, where it is often critical for the operation of the most diverse systems. The need to develop systems that save electricity has been mentioned earlier. In fact, software can be developed under the "greener" nomenclature, with the main concern of minimizing damage to the environment, starting with energy savings. In fact, the management of the most diverse devices can be made more versatile if the software programming obeys the principles of environmental sustainability. Just as it is possible to select a fuel management that is more favorable to the environment in a car, avoiding high torsion of combustion engines through proper management of the gearbox, or even the management of accelerations and speed when driving electric vehicles, the same can be done with systems very dependent on electricity, to promote a more rational use of energy. It is in the management of energy consumption that the software has found a greater field of application to promote electrical sustainability. However, certainly in the future, with the help of Artificial Intelligence, the devices that we use in the most diverse situations will recommend solutions that make our options more compatible with the needs of the environment, helping us reflect on the options that we take and trying to guide all those who are more sensitive to environmental issues in the most adequate way so that their practices are effectively consistent with their environmental sensitivity.

In addition to energy consumption, paper consumption is also extremely harmful for the environment, as it involves the cultivation and cutting of huge amounts of eucalyptus, which are transformed into cellulose, and then processed until reaching the various forms of paper we consume. Paper recycling has increased significantly, but it cannot cover all market needs. In 2021, 71.4% of all paper and board consumed in Europe was recycled and the European Union aims to recycle 76% of the paper used by 2030 [75], which is an ambitious goal. If applications such as packaging, toilet paper, cardboard and other less relevant applications can perfectly be carried out using recycled paper, there is still a tendency for most service companies, including official institutions that issue documents, to continue preferring non-recycled paper due to its whiteness level. Thus, an immense number of trees continue to be felled to be transformed into paper pulp and, subsequently, high-quality paper.

The problem of felling trees is not only caused by the excessive consumption of paper. In fact, there are many documents that could circulate digitally and, to the detriment of that, are printed, generating another problem: the consumption of inks, which makes the recycling process more difficult, making it more expensive and slower. Automation, together with the internet of things, can play a key role in minimizing the environmental impact that these processes can have. In fact, most commercial transactions between companies or between companies and individuals generate the issuance of documents. Most bank transactions generate documents. Tickets for public transportation, shows, and others generate documents. And there are many other situations where there is a need to issue documents. However, most of these documents can currently circulate digitally, producing the same effect, or even a more significant effect. In fact, the acquisition of any good by an individual from a company is normally accompanied by a document, which can be used for fiscal control by the customer and by the company. However, if the services were digitized, this document can be issued, sent to the client and the Government in digital format, produce all the necessary effects and be accessible at any time, without the need to consume paper and printing ink. In more developed countries, service companies are using consumer acceptance policies for these digital practices, not only with the aim of benefiting the environment, but because this also represents significant savings in operating costs. Banks, energy distribution companies, telecommunications operators, credit institutions, among many others, are encouraging customers to choose to receive documents in digital format, offering discounts, which result from the savings of resources they now have in their operations. This is just one example of how digitization through the automation of processes in the service sector, based on technological platforms, can help environmental policies, even though it is driven by a clear focus on cost reduction.

Traditionally, the acquisition of a household appliance or any other device is accompanied by user manuals in different languages, depending on the markets where it is presumed to be sold. This represents a significant amount of paper and ink. Given that the use of the internet of things is highly disseminated, mainly in more developed or developing countries, any customer's access to information regarding the installation and operation of any device can be provided through a simple QR code on the packaging, that will allow direct access to all information about the product. There is still a long way to go in this regard, as this situation is not always welcomed in the best way by older people who are less sensitive to the environment. Undoubtedly, the first step will be to educate people to use the internet of things. Then, the platforms must be designed in such a way that their use is as user-friendly and reliable as possible, creating customer trust. Finally, initial incentives for the use of documents in digital format are a fundamental step towards expanding the customer base in this mode of use, which tends to grow naturally because of the influence of family and friends. The digitization of the issuance of tickets corresponding to air flights met some initial resistance but, from the moment when the number of service desks began to be reduced and travelers began facing longer service times, led to many starting their own digitization process, starting to use the tickets in digital format on their smartphone. This is the long path that most services will have to take in order to convince people who are less sensitive to the environment that the digital path is the path we all have to take for the sake of the environment.

Health systems can be generators of countless documents whose circulation can present some complexity. However, the creation of reliable platforms designed from the perspective of the necessary information flows, could eliminate the need to print medication prescriptions or exams, by simply sending a message that can be read by pharmacies, hospitals, or clinics. In this way, paper consumption and printing of documents that would hardly be of any use at the end of the usual information cycle is avoided. On the other hand, information continues to flow between the doctor and health care providers, pharmacies, or clinics. All these stakeholders can also make information available in digital form, which can be easily accessed in a free format accessible to most devices used by patients and by those who prescribe medications, tests, or treatments. The entire circuit of information, from the prescription to the reception of the results of an exam or the invoice for the medication purchased, can be circulated in a digital format, reducing the sending time between the different stages of the process and not creating any embarrassment for the patient, provided that the same is minimally familiar with the use of smartphones or computers. Obviously, there are assumptions that need to be met. The older population or those with less appetite for new technologies will experience greater difficulty in keeping up with these processes. However, the advantages for most stakeholders are so evident, both in organizational terms and in terms of economic and environmental sustainability, that the most evolved societies are moving irreversibly in this direction. In more developed countries, public health entities already issue prescriptions in this way, gathering all the patient's information in powerful databases, even facilitating the change of doctor in case of need, without losing any type of information regarding the patient. The advantages at all levels are evident, and this type of process automation in the health area brings added benefits to the environment.

Considering production scheduling and industrial management as a service, it is still common in many companies to use appreciable amounts of documents that accompany the production of parts and components, or even the assembly of more complex systems. The electronics industry and the automotive industry have set an excellent example in this respect, digitizing the process of putting the product into production, monitoring, and controlling it. The use of QR codes from its initial production phase, the follow-up service being entirely carried out digitally, avoiding the circulation of paper documents, which, when not digitized, often get dirty, making consultation unfeasible, or are lost, causing confusion in its tracking, are very helpful tools in some manufacturing sectors. The digitization of information allows for greater centralization of information, promoting greater speed in recognizing the different states of the product and providing the appropriate information necessary for tracking or completing the product. The use of monitors, optical readers, QR codes and effective information systems has fostered much more efficient production flows, with clear benefits for the environment as well.

In this regard, some architecture topics that commonly use the names of "Industry 4.0" or "Smart Manufacturing", have contributed significantly to what was mentioned above. The centralization of information, as well as the command and control of equipment, is now carried out without any need for support in paper format, which facilitates the workflow and substantially helps the environment. Automation is behind all these processes, being directly and indirectly responsible for processes that are now much more environmentally friendly.

The development of digital services has converged into a new business area, e-commerce. This business model, which has grown significantly over the past two decades, has proven to be environmentally friendly, although there are also some associated problems. In positive terms, the form of purchase should be highlighted, avoiding the creation, printing, and distribution of catalogs, with the consequent consumption of paper and ink. In addition, consumers can purchase their goods digitally, without having to go to different stores, where they may not have the product they want or the size they need. Supply becomes centralized, so that logistics costs by distribution companies can be minimized. In fact, if the entire business model was based only on centralized warehouses from which all orders for end customers would leave, distribution to countless retail stores would be avoided, with the consequent savings in trips in distribution between central warehouses and retail stores. If e-commerce generates a greater number of trips between the central warehouses and the final consumer, the savings are obvious since it replaces two routes: central warehouse to distributor, and from the customer to the distributor. The optimization of routes by logistics companies can also help distribution avoid a greater number of emissions, which will necessarily be lower than if all consumers needed to travel to one or even several warehouses to purchase the product they need. This optimization could be the same to the optimization that the consumer could carry out when purchasing several products in a single trip. The optimization of deliveries by logistics companies may even evolve in the future to another type of service, through autonomous robots for delivering orders, a subject discussed later in this chapter. In addition, the entire bureaucratic process behind the business is also digitalized, avoiding the circulation of printed paper, with consequent savings for the environment in terms of paper and ink. As a negative point, it can be mentioned that each product needs to be properly packaged, which consumes cardboard and plastic, thus being harmful to the environment. Obviously, recycled cardboard and plastic can be used to minimize the environmental impact, but these practices are also common in traditional trade. In fact, packaging continues to be a global problem in terms of wasting resources, which needs to be addressed urgently. The ability to make products and processes more environmentally friendly will necessarily have to evolve in the packaging sector, where there are still important savings that can be made in favor of the environment. In any case, it can be seen that the services sector can take enormous advantage of electronic platforms, developed software, and process automation. Once again, automation plays a key role in saving resources and, in this case, also in the emission of greenhouse gases, with significant gains for the environment.

However, even though it looks like a very environmentally friendly solution, it brings some challenges that should be considered. In addition to some constraints related to the availability of who delivers and who receives (drop-off time window), this system has different attributes when inserted in an urban environment or in an environment with less population density. The rationality and optimization of trips within an urban environment allows for faster deliveries, which better satisfy the consumer, while in less populated environments, deliveries need to be grouped so that delivery is optimized, depending on the logistic channels established. Uncertainty regarding the volume of deliveries to be made each day leads logistics companies to subcontract last-mile delivery services to contractors. This task, being somewhat stressful and usually poorly paid, is only interesting for relatively young people with no perfectly defined career prospects. Additionally, working conditions can be even more stressful if one considers that those who carry out last-mile delivery are exposed to road traffic and weather conditions specific to each region and time of year. Given these working conditions are not the most satisfactory, companies may soon find it difficult to find people available for this type of task, what is known as Autonomous Robot-driven Delivery [76] is being developed, as shown in Fig. 3.2.

Although autonomous robots seem like an alternative that can bring clear benefits to the environment, and there are developments that predict that this reality could easily be extended to other less developed regions, even in developed or developing countries there are still many barriers to overcome, namely social acceptance (assumes the dismissal of people), adequate infrastructure, fleets compatible with the volume of deliveries, dynamic routing, traffic constraints for some robots considered, downtime for recharging other types of robots, among others. The lack of contact between humans in deliveries also needs to be properly managed, to provide experiences that are perfectly accepted by customers. In addition to ground delivery, there are also studies aimed at deliveries by air, using autonomous drones. These efforts could also promote an integrated logistics service for hybrid deliveries, linking transport carried out by trucks, drones, and robots.

Automation and robotics have been applied with increasing frequency in other service sectors, creating a mixed sensation on the part of the customer: the attraction for the novelty it represents, and the fear that robots will take the work away from humans, creating a real social issue. In pharmacies, robots are something more and more common, facilitating the task of those who serve, minimizing the waiting time and reducing the risks of error whenever possible in human service. These robots follow the principle of any automatic service, identifying the necessary product, searching for its location in the warehouse duly referenced, picking up the product in the desired quantity and delivering it together at the counter to the operator who enter

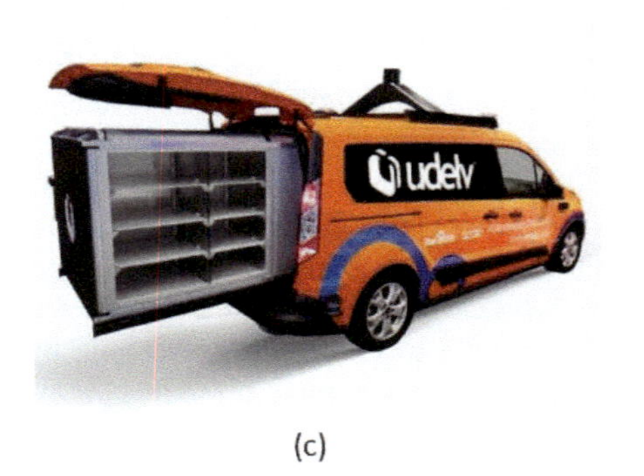

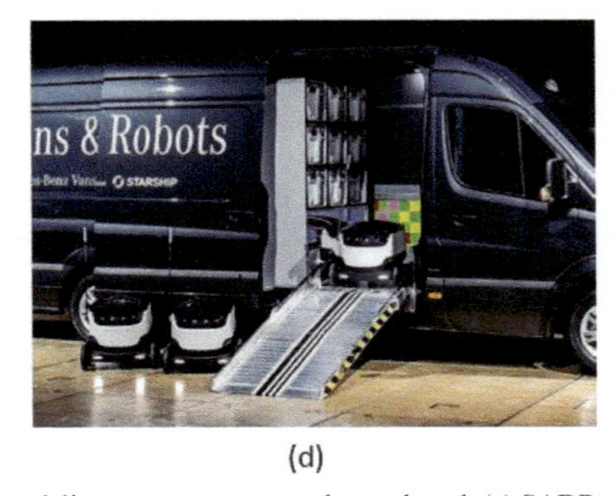

Figure 3.2. Different ground autonomous robot-driven delivery concepts currently employed: (a) SADR (source: Starship Technologies), (b) RADR (source: Nuru), (c) ADV (source: Udelv), (d) Mothership (source: Mercedes - Benz) [76].

the data into the computer. In hotels, robotic service is already a reality, especially in more developed countries such as Japan, the United States, and some European countries [77]. Even restaurants have adopted robots, as is the case of restaurant chains installed in China under the name of Haidilao and Spyce [78], which are equipped with automated kitchens with robot chefs to increase efficiency. However, no clear benefits for the environment with this type of service automation have been reposted so far.

Through what has been described in this chapter, it can be seen that automation and robotics, together with technological evolution in terms of computational capacity, artificial intelligence, smarter and user-friendly mobile devices, more user-friendly software and the expansion of the internet of things in the world's population, has contributed significantly to a global improvement of the environment, in terms of saving natural resources, saving energy, reducing emissions harmful to the environment, and less waste produced, whether this is recyclable or just for deposit in landfills. Since this development is based on technological evolution, it should also be noted that technology also has some harmful effects on the environment, as this same evolution makes obsolete many devices that contain extremely harmful products for the environment. For this reason, and despite the efforts made so far, which should be praised and encouraged, they are still insufficient to counteract the effect caused by the consumerist spirit that still invades the minds of a large part of the citizens, mainly from more developed countries and with better wages. This trend needs to be reversed, but the world still seems to be far from reaching that turning point.

References

[1] Gomez-Baggethun, E. (2022). Rethinking work for a just and sustainable future. Ecological Economics, 200: 107506. doi: 10.1016/j.ecolecon.2022.107506.

[2] Santos, R. F. L., Silva, F. J. G., Gouveia, R. M., Campilho, R. D. S. G., Pereira, M. T. and Ferreira, L. P. (2018). The Improvement of an APEX Machine involved in the Tire Manufacturing Process. Procedia Manufacturing, 17: 571–578. doi: 10.1016/j.promfg.2018.10.098.

[3] Zhang, Q., Zhang, F. and Mai, Q. (2022). Robot adoption and green productivity: Curse or Boon. Sustainable Production and Consumption 34 (2022) 1–11. doi: 10.1016/j.spc.2022.08.025.

[4] Liang, T., Zhang, Y. J. and Qiang,W. (2022). Does technological innovation benefit energy firms' environmental performance? The moderating effect of government subsidies and media coverage. Technological Forecasting and Social Change, 180: 121728. doi: 10.1016/j.techfore.2022.121728.

[5] Yang, L. and Li, Z. (2017). Technology advance and the carbon dioxide emission in China—empirical research based on the rebound effect. Energy Policy, 101: 150–161. doi: 10.1016/j.enpol.2016.11.020.

[6] Brännlund, R., Ghalwash, T. and Nordström, J. (2007). Increased energy efficiency and the rebound effect: effects on consumption and emissions, Energy Economics, 29: 1–7. doi: 10.1016/j.eneco.2005.09.003.

[7] Binswanger, M. (2001). Technological progress and sustainable development: What about the rebound effect? Ecological Economics, 36: 119–132. doi: 10.1016/S0921-8009(00)00214-7.

[8] Araújo, W. F. S., Silva, F. J. G., Campilho, R. D. S. G. and Matos, J. A. (2017). Manufacturing cushions and suspension mats for vehicle seats: A novel cell concept, The International Journal of Advanced Manufacturing Technology, 90: 1539–1545. doi: 10.1007/s00170-016-9475-6.

[9] Nazareno, L. and Schiff, D. S. (2021). The impact of automation and artificial intelligence on worker well-being, Technology in Society, 67: 101679. doi: 10.1016/j.techsoc.2021.101679.

[10] Manyika, J., Lund, S., Chui, M., Bughin, J., Woetzel, J., Batra, P., Ko, R. and Sanghvi, S. (2017). What the future of work will mean for jobs, skills, and wages: Jobs lost, jobs gained. McKinskey Global Institute. URL: https://www.mckinsey.com/featured-insights/future-of-work/jobs-lost-jobs-gained-what-the-future-of-work-will-mean-for-jobs-skills-and-wages, (Accessed on December 8th, 2022).

[11] Ljubica Nedelkoska and Glenda Quintini. (2018). Automation, Skills Use and Training. URL: https://doi.org/10.1787/2e2f4eea-en, (Accessed on December 8th, 2022).

[12] Gan, J., Liu, L., Qiao, G. and Zhang, Q. (2023). The role of robot adoption in green innovation: Evidence from China, Economic Modelling, 119: 106128. doi: 10.1016/j.econmod.2022.106128.

[13] Bonfiglioli, A., Crinò, R., Fadinger, H. and Gancia, G. (2020). Robot Imports and Firm-Level Outcomes. CESifo Working Paper, p. 8741. URL: chrome-extension: //efaidnbmnnnibpcajpcglclefindmkaj/https://www.econstor.eu/bitstream/10419/229559/1/cesifo1_wp8741.pdf, (Accessed on December 9th, 2022).

[14] Acemoglu, D. and Pascual, R. (2018a). The race between man and machine: implications of technology for growth, factor shares, and employment. American Economic Review, 108: 1488–1542. doi: 10.1257/aer.20160696.

[15] Acemoglu, D. and Pascual, R. (2018b). Low-skill and high-skill automation Journal of Human Capital. University of Chicago Press, 12(2): 204–232. URL: chrome-extension://efaidnbmnnnibpcajpcglclefindmkaj/https://economics.mit.edu/sites/default/files/publications/Low-Skill%20and%20High-Skill%20Automation.pdf, (Accessed on December 9th, 2022).

[16] Sabarwal, A., Kumar, K. and Singh, R. P. (2018). Hazardous effects of chemical pesticides on human health–Cancer and other associated disorders, Environmental Toxicology and Pharmacology, 63: 103–114. doi: 10.1016/j.etap.2018.08.018.

[17] Grand View Research (2022). Organic Food and Beverages Market Size, Share & Trends Analysis, Report By Product. URL: https://www.grandviewresearch.com/industry-analysis/organic-foods-beverages-market, (Accessed on December 9th, 2022).
[18] Husaini, A. M. and Tuteja, N. (2013). Biotech crops: Imperative for achieving the Millenium Development Goals and sustainability of agriculture in the climate change era, GM crops & food, 4(1): 1–9. doi: 10.4161/gmcr.22748.
[19] Husaini, A. M. and Sohail, M. (2022). Robotics-assisted, organic agricultural biotechnology based environment-friendly healthy food option: Beyond the binary of GM versus Organic crops. Journal of Biotechnology, 361: 41–48. doi: 10.1016/j.jbiotec.2022.11.018.
[20] Zikeli, S. and Gruber, S. (2017). Reduced tillage and no-till in organic farming systems, Germany—Status quo, potentials and challenges. Agriculture, 7(4): 35. doi: 10.3390/agriculture7040035.
[21] Jha, K., Doshi, A., Patel, P. and Shah, M. (2019). A comprehensive review on automation in agriculture using artificial intelligence, Artificial Intelligence in Agriculture, 2: 1–12. doi: 10.1016/j.aiia.2019.05.004.
[22] Lemmon, H. (1986). Comax: An expert system for cotton crop management. Science, 233(4759): 29–33. doi: 10.1126/science.233.4759.29.
[23] Roach, J., Virkar, R., Drake, C. and Weaver, M. (1987). An expert system for helping apple growers. Computers and Electronics in Agriculture, 2(2): 97–108. doi: 10.1016/0168-1699(87)90020-2.
[24] Stone, N. D. and Toman, T. W. (1989). A dynamically linked expert-database system for decision support in Texas cotton production. Computers and Electronics in Agriculture, 4(2): 139–148. doi: 10.1016/0168-1699(89)90031-8.
[25] Batchelor, W. D., McClendon, R. W., Adams, D. B. and Jones, J. W. (1989). Evaluation of SMARTSOY: An expert simulation system for insect pest management. Agricultural Systems, 31(1): 67–81. doi: 10.1016/0308-521X(89)90013-9.
[26] Robinson, C. and Mort, N. (1997). A neural network system for the protection of citrus crops from frost damage. Computers and Electronics in Agriculture, 16(3): 177–187. doi: 10.1016/S0168-1699(96)00037-3.
[27] Maier, H. R. and Dandy, G. C. (2000). Neural networks for the prediction and forecasting of water resources variables: A review of modeling issues and applications. Environmental Modeling & Software, 15(1): 101–124. doi: 10.1016/S1364-8152(99)00007-9.
[28] Gliever, C. and Slaughter, D. C. (2001). Crop verses weed recognition with artificial neural networks. ASAE annual meeting. Paper. 01-3104 (2001), 1–12. doi: 10.13031/2013.7425.
[29] Song, H. and He, Y. (2005). Crop nutrition diagnosis expert system based on artificial neural networks. Third International Conference on Information Technology and Applications (ICITA'05), Sydney, NSW, 1: 357–362, 2005. doi: 10.1109/ICITA.2005.108.
[30] Aitkenhead, M. J., Dalgetty, I. A., Mullins, C. E., McDonald, A. J. S. and Strachan, N. J. C. (2003). Weed and crop discrimination using image analysis and artificial intelligence methods. Computers and Electronics in Agriculture, 39(3): 157–171. doi: 10.1016/S0168-1699(03)00076-0.
[31] Singh, R. K. and Prajneshu. (2008). Artificial neural network methodology for modelling and forecasting maize crop yield. Agricultural Economics Research Review, 21: 5–10. doi: 10.22004/ag.econ.47354.
[32] Hinnell, A. C., Lazarovitch, N., Furman, A., Poulton, M. and Warrick, A. W. (2010). Neuro-drip: Estimation of subsurface wetting patterns for drip irrigation using neural networks. Irrigation Science 28: 535–544. doi: 10.1007/s00271-010-0214-8.
[33] Arif, C., Mizoguchi, M., Setiawan, B. I. and Doi, R. (2012). Estimation of soil moisture in paddy field using Artificial Neural Networks. International Journal of Advanced Research in Artificial Intelligence, 1(1): 17–21. doi: 10.48550/arXiv.1303.1868.
[34] Keshtgari, M. and Deljoo, A. (2012). A wireless sensor network solution for precision agriculture based on ZigBee technology. Wireless Sensor Network. 4(1): 1–7. doi: 10.4236/wsn.2012.41004.
[35] Ravichandran, G. and Koteshwari, R. S. (2016). Agricultural crop predictor and advisor using ANN for smartphones. 2016 International Conference on Emerging Trends in Engineering. Technology and Science (ICETETS), pp. 1–6. doi: 10.1109/ICETETS.2016.7603053.
[36] Nema, M. K., Khare, D. and Chandniha, S. K. (2017). Application of artificial intelligence to estimate the reference evapotranspiration in sub-humid Doon valley. Applied Water Science, 7: 3903–3910. doi: 10.1007/s13201-017-0543-3.

[37] Ingale, H. T. and Kasat, N. N. (2012). Automated irrigation system. International Journal of Engineering Research and Development, 4(11): 51–54. doi: N/A.
[38] Roopaei, M., Rad, P. and Choo, K. K. R. (2017). Cloud of things in smart agriculture: intelligent irrigation monitoring by thermal imaging. IEEE Cloud Computing, 4(1): 10–15. doi: 10.1109/MCC.2017.5.
[39] Vadivambal, R. and Jayas, D. (2011). Applications of thermal imaging in agriculture and food industry—A review. Food and Bioprocess Technology, 4(2): 186–199, doi:10.1007/s11947-010-0333-5.
[40] Yong, W., Shuaishuai, L., Li, L., Minzan, L., Arvanitis, K. G., Georgieva, C. and Sigrimis, N. (2018). Smart sensors from ground to cloud and web intelligence. IFAC-Papers OnLine 51(17): 31–38. doi: 10.1016/j.ifacol.2018.08.057.
[41] Tian, H., Wang, T., Liu, Y., Qiao, X. and Li, Y. (2020). Computer vision technology in agricultural automation—A review, Information Processing in Agriculture, 7: 1–19. doi: 10.1016/j.inpa.2019.09.006.
[42] Gil, G., Casagrande, D. E., Cortés, L. P. and Verschae, R. (2023). Why the low adoption of robotics in the farms? Challenges for the establishment of commercial agricultural robots. Smart Agricultural Technology, 3. doi: 10.1016/j.atech.2022.100069.
[43] Poore, J. and Nemecek, T. (2018). Reducing food's environmental impacts through producers and consumers. Science, 360: 6392. doi: 10.1126/science.aaq0216.
[44] Melián-Navarro, A., Molina-Martínez, J. M., Rodríguez-Díaz, J. A. and Ruiz-Canales, A. (2017). Performance indicators to assess the implementation of automationin golf courses located in Southeast Spain. Agricultural Water Management, 183: 35–40. doi: 10.1016/j.agwat.2016.11.017.
[45] UN Environment Programme, 2022 Global Status Report for Buildings and Construction. URL: file:///C:/Users/asus/Desktop/Building_Construction_2022.pdf, (Accessed on December 8th, 2022).
[46] Lin, S.-H., Zhang, H., Li, J.-H., Ye, C.-Z. and Hsieh, J.-C. (2022). Evaluating smart office buildings from a sustainability perspective: A model of hybrid multi-attribute decision-making. Technology in Society, 68: 101824. doi: 10.1016/j.techsoc.2021.101824.
[47] European Committee. (2009). ICT for a Low Carbon Economy- Smart Buildings, European Commission, Belgium.
[48] Fetsis, P. The LIFE Programme – Over 20 years improving sustainability in the built environment in the EU. Procedia Environmental Sciences, 38: 913–918. doi: 10.1016/j.proenv.2017.03.179.
[49] Bertim. (2016). Building Energy Renovation through Timber Prefabricated Modules. http://www.bertim.eu/, (Accessed on December 19th, 2022).
[50] Pan, M., Linner, T., Pan, W., Cheng, H. and Bock, T. (2018). A framework of indicators for assessing construction automation and robotics in the sustainability context. Journal of Cleaner Production, 182: 82–95. doi: 10.1016/j.jclepro.2018.02.053.
[51] Salahi, N. and Jafari, M. A. (2016). Energy-performance as a driver for optimal production planning. Applied Energy, 174: 88–100. doi: 10.1016/j.apenergy.2016.04.085.
[52] International Federation of Robotics. (2022). World Robotics Report: "All-Time High" with Half a Million Robots Installed in one Year,URL: https://ifr.org/ifr-press-releases/news/wr-report-all-time-high-with-half-a-million-robots-installed, (Accessed on December 26th, 2022).
[53] Fysikopoulos, A., Anagnostakis, D., Salonitis, K. and Chryssolouris, G. (2012). An empirical study of the energy consumption in automotive assembly. Procedia CIRP, 3: 477–482. doi: 10.1016/j.procir.2012.07.082.
[54] Lofving, M., Almstrom, P., Jarebrant, C., Wadman, B. and Widfeldt, M. (2018). Evaluation of flexible automation for small batch production. Procedia Manufacturing, 25: 177–184. doi: 10.1016/j.promfg.2018.06.072.
[55] Araújo, W. F. S., Silva, F. J. G., Campilho, R. D. S. G. and Matos, J. A. (2017). Manufacturing cushions and suspension mats for vehicle seats: A novel cell concept. International Journal of Advanced Manufacturing Technology, 90: 1539–1545. doi: 10.1007/s00170-016-9475-6.
[56] Costa, R. J. S., Silva, F. J. G. and Campilho, R. D. S. G. (2017). A novel concept of agile assembly machine for sets applied in the auto-motive industry. International Journal of Advanced Manufacturing Technology, 91: 4043–4054. doi: 10.1007/s00170-017-0109-4.

[57] Moreira, B. M. D. N., Gouveia, R. M., Silva, F. J. G. and Campilho, R. D. S. G. (2017). A novel concept of production and assembly processes integration. Procedia Manufacturing, 11: 1385–1395. doi: 10.1016/j.promfg.2017.07.268.

[58] Costa, M. J. R., Gouveia, R. M., Silva, F. J. G. and Campilho, R. D. S. G. (2018). How to solve quality problems by advanced fully-automated manufacturing systems. International Journal of Advanced Manufacturing Technology, 94: 3041–3063. doi: 10.1007/s00170-017-0158-8.

[59] Silva, F. J. G., Swertvaegher, G., Campilho, R. D. S. G., Ferreira, L. P. and Sá, J. C. (2020). Robotized solution for handling complex automotive parts in inspection and packing. Procedia Manufacturing, 51: 156–163. doi: 10.1016/j.promfg.2020.10.023.

[60] Figueiredo, D., Silva, F. J. G., Campilho, R. D. S. G., Silva, A., Pimentel, C. and Matias, J. C. O. (2020). A new concept of automated manufacturing process for wire rope terminals. Procedia Manufacturing, 51: 431–437. doi: 10.1016/j.promfg.2020.10.061.

[61] Santos, P. M. M., Campilho, R. D. S. G. and Silva, F. J. G. (2021). A new concept of full-automated equipment for the manufacture of shirt collars and cuffs. Robotics and Computer-Integrated Manufacturing, 67: 102023. doi: 10.1016/j.rcim.2020.102023.

[62] Silva, F. J. G., Soares, M. R., Ferreira, L. P., Alves, A. C., Brito, M., Campilho, R. D. S. G. and Sousa, V. F. C. (2021). A Novel Automated System for the Handling of Car Seat Wires on Plastic Over-Injection Molding Machines, 9: 141. doi: 10.3390/machines9080141.

[63] Sousa, V. F. C., Silva, F. J. G., Campilho, R. D. S. G., Pinto, A. G., Ferreira, L. P. and Martins, N. (2022). Developing a novel fully automated concept to produce bowden cables for the automotive industry. Machines, 10: 290. doi: 10.3390/machines10050290.

[64] Sousa, V., Silva, F. J. G., Fecheira, J. S., Campilho, R. D. S. G. and Vandermeulen, V. (2020). A Novel modular design of an equipment to produce "T"-profiles by laser welding. Procedia Manufacturing, 51: 446–453. doi: 10.1016/j.promfg.2020.10.063.

[65] Ogbemhe, J., Mpofu, K. and Tlale, N. S. (2017). Achieving sustainability in manufacturing using robotic methodologies. Procedia Manufacturing, 8: 440–446. doi: 10.1016/j.promfg.2017.02.056.

[66] Castro, A. F., Silva, M. F. and Silva, F. J. G. (2017). Designing a robotic welding cell for bus body frame using a sustainable way. Procedia Manufacturing, 11: 207–214. doi: 10.1016/j.promfg.2017.07.225.

[67] Kim, J. S., Son, Y. T., Cho, H. S. and Koh, K. I. (1995). A robust method for vision-based seam tracking in robotic arc welding. Proceedings of Tenth International Symposium on Intelligent Control, 363–368. doi: 10.1109/ISIC.1995.525084.

[68] Wang, K., Zhang, J., Zhao, B. and Tang, D. (2010). Research on Control Technology of Trajectory Tracking for Robotic Welding, International Conference on Computational Intelligence and Software Engineering, 1–4, doi: 10.1109/CISE.2010.5676764.

[69] Vergnano, A., Thorstensson, C., Lennartson, B., Falkman, P., Pellicciari, M., Leali, F. and Biller, S. (2012). Modeling and optimization of energy consumption in cooperative multi-robot systems. IEEE Transactions on Automation Science and Engineering, 9(2): 423–428. doi: 10.1109/TASE.2011.2182509.

[70] Wang, E.-Z., Lee, C.-C. and Li, Y. (2022). Assessing the impact of industrial robots on manufacturing energy intensity in 38 countries. Energy Economics, 105(2022): 105748. doi: 10.1016/j.eneco.2021.105748.

[71] Zhou, B. and Wu, Q. (2020). Decomposition-based bi-objective optimization for sustainable robotic assembly line balancing problems. Journal of Manufacturing Processes, 55: 30–43. doi: 10.1016/j.jmsy.2020.02.005.

[72] Merlo, F., Vazzoler, G. and Berselli, G. (2023). Eco-programming of industrial robots for sustainable manufacturing via dynamic time scaling of trajectories. Robotics and Computer–Integrated Manufacturing, 79: 102420. doi: 10.1016/j.rcim.2022.102420.

[73] Junior, A. N., de Oliveira, M. C. and Helleno, A. L. 2018. Sustainability evaluation model for manufacturing systems based on the correlation between triple bottom line dimensions and balanced scorecard perspectives. Journal of Cleaner Production, 190: 84–93. doi: 10.1016/j.jclepro.2018.04.136.

[74] Chen, X., Despeisse, M. and Johansson, B. (2020). Environmental sustainability of digitalization in manufacturing: A review. Sustainability, 12(24): 10298. doi: 10.3390/su122410298.

[75] Recycling Magazine. (2022). Paper value chain reached a 71.4% recycling rate. URL: https://www.recycling-magazine.com/2022/09/09/paper-value-chain-reached-a-71-4-recycling-rate/. (Accessed on December 29th, 2022).
[76] Srinivas, S., Ramachandiran, S. and Rajendran, S. (2022). Autonomous robot-driven deliveries: A review of recent developments and future directions. Transportation Research Part E, 165, 102834. doi: 10.1016/j.tre.2022.102834.
[77] Orea-Giner, A., Fuentes-Moraleda, L., Villacé-Molinero, T., Muñoz-Mazón, A. and Calero-Sanz, J. (2022). Does the implementation of robots in hotels influence the overall TripAdvisor rating? A Text Mining Analysis from the Industry 5.0 Approach, Tourism Management, 93: 104586. doi: 10.1016/j.tourman.2022.104586.
[78] Wang, Y., Kang, Q., Zhou, S., Dong, Y. and Liu, J. (2022). The impact of service robots in retail: Exploring the effect of novelty priming on consumer behavior. Journal of Retailing and Consumer Services, 68: 103002. doi: 10.1016/j.jretconser.2022.103002.

Chapter 4

Automation Applied to Manufacturing Systems

4.1 Contextualization, green manufacturing, and remanufacturing

Although we all contribute with our daily activity, whether professional or domestic, to the emission of greenhouse gases, one-fifth of the world's carbon emissions and 54% of the energy consumption come from the manufacturing sector and related activities [1]. Additionally, supply chains related to manufacturing activities produce about 11 times more emissions than the production operations themselves. These facts are very worrying, although most of the larger companies have plans to decarbonize their production activities. However, the need to act quickly, and in the most diverse ways, becomes evident. To overcome this problem, there are proposals for studying and recording data on the ecological footprint left by each product during its life cycle, using a common base called CO_2-equivalent or carbon-equivalent, thus allowing manufacturers to identify which are the most critical phases and try reducing them. This should start by the project of the products, which design and material options will be impacted by the desired improvements for the environment.

Pollution generated by industry can be classified according to five aspects: air, water, soil, noise, and light. Of all these groups, according to the World Health Organization [2], air pollution is responsible for about 7 million deaths worldwide, this impact was measured by the number of people who die from respiratory or related diseases resulting from inhaling air polluted with toxic products. However, the direct and/or indirect impact generated by other types of pollution is much more difficult to assess, but the World Health Organization [2] estimates that water pollution is at the base of about 500,000 premature deaths per year due to diarrhea, while the immense amount of plastic that is dumped into the sea may be the cause of a few million deaths of children, animals, and fish. Although all types of pollution are of concern, there is no doubt that environmental air pollution is the one that has the greatest direct impact on human health and generates the greatest concern. The number of deaths due to ambient air pollution includes deaths from pollution generated by industry and household air pollution, which result in illnesses that underlie deaths due to strokes, heart diseases, lung cancer, acute and chronic respiratory diseases. The major outdoor pollution sources include residential energy for heating and cooking, vehicles provided with combustion engines, electrical power generation, agriculture, waste incineration, and industry. In order to mitigate the problem of ambient air pollution, it is urgent that policies be established that promote the transition to sustainable land use, cleaner household energy and transport, energy-efficient housing, electrical power generation, reduced industry effluents, and efficient municipal waste management.

One of the biggest concerns in terms of pollution is the release of carbon. Life on Earth, in natural terms, induces the release of carbon by animals and plants. Plants have their own carbon release and absorption cycle, releasing it during the night and absorbing it during the day. As far as animals are concerned, they essentially breathe in oxygen and breathe out carbon dioxide throughout the day. However, the Earth has its own capacity to regenerate this carbon dioxide, in a cycle that is, by nature and due to nature, balanced.

The exploitation and intensive use of fossil products has unbalanced the entire balance that nature had, since the amount of carbon released to nature greatly exceeds the capacity that the oceans and living matter have to reabsorb that carbon, and that is only 40% of the amount that is currently generated, while the rest remains in the atmosphere, creating what is known as global warming. The remaining carbon, not being reabsorbed, causes the abovementioned problems. This problem was clearly identified several decades ago, but policies to mitigate the problem are slow to be implemented, essentially due to economic interests. The way we move around has been based on combustion engines for about a century, which release carbon into the environment. The industry generated, and still generates, electrical energy based on coal, seriously polluting the environment through strong carbon emissions into the environment. Even other forms of electricity generation depend on petroleum products. The polymers, glues and adhesives industries are extremely dependent on petroleum products. This exacerbated consumption of petroleum products created dependence on some countries, and clear benefits/profits on others, giving rise to lobbies that were installed and which, due to the high profits achieved through a price policy regulated by the group themselves, prevented for a long time that other approaches to locomotion had been developed earlier. The generation of electricity by renewable means brought a breath of fresh air in the generation of electric energy, which only started to be democratized in the second decade of the 21st century. Finally, and facing the abyss, more courageous goals have been established for the energy transition, and it is foreseeable that the war induced by Russia in 2022 will contribute positively to the acceleration of even more ambitious goals for this energy transition, clearly showing that the energy dependency on the supply of fossil fuels is a serious problem, due to the pollution they generate. Moreover, industry should reduce its greenhouse gases production by 80% by 2050, allowing to avoid a temperature rising over 2°C. Although still in many countries their economic interests prevail over the energy transition, there finally appears to be an effective need to conduct this energy transition and change industrial paradigms to avoid more serious geopolitical problems, which suggests being much more important for those who are responsible for making decision policies, than the preservation of the environment and health of their populations. It is also a fact that the energy transition is not something simple to execute, as it brings immense implications to our current way of living, and especially to the entire way in which business and taxes are instituted. In any case, there appears to be a consensus now that this step must be taken, and there are goals to take this step, but we should wait to confirm whether this step will be taken in a timely manner.

In industry, energy production ranks top in terms of pollution, contributing around 30% of greenhouse gas emissions (mainly carbon dioxide and methane), while at the United Nations Climate Change Conference held at the SEC Center in Glasgow, Scotland, UK, from 31st October to 13th November 2021, coal was highlighted as the main cause of the climate change currently occurring, and coal is still used in many countries as one of the main resources for electrical power generation.

Transport, which is partly linked to logistical operations essential to the movement of materials, semi-products and finished products manufactured by industry, is also a problem for the environment, accounting for around 20% of greenhouse gas emissions. These transports correspond roughly to 60% for citizens traveling in their daily activities, while around 40% correspond to commercial and industrial activities. Most of these vehicles still move based on internal combustion engines, with oil-based fuels. In 2022, the average consumption of oil in world terms was around 100 million barrels per day, while in 2006 it was around 85 million barrels per day, allowing to observe that consumption continues to increase, despite all political commitments made around the environment. The extraction, refining and transport of this oil entails risks for the environment, as possible leaks in pipelines and accidents with oil tankers, among others, can cause spills with very harmful consequences for the environment. Therefore, the paradigm shifts in vehicle locomotion, with a transition to electric mobility or hydrogen cells use, could be an excellent solution for the environment, considering that electrical energy is produced by renewable means and does not generate more greenhouse gases.

The fashion industry is responsible for around 10% of greenhouse gas emissions, and there are still growth prospects for this industrial sector. In addition to all this, the most common fashion industry is carried out mainly in less developed or developing countries, where the energy production is carried out essentially based on coal. Therefore, this is a sector that should deserve more attention in the short term, allowing this type of industry to adopt strategies to reduce energy consumption and put pressure on the authorities in each country to increase the share of electricity produced based on renewable means.

The remaining emissions to reach 100% concern essentially those produced by the agricultural sector and by the food industry. Concerning the food industry, it was estimated that around 2 million tons of food per year are spoiled in the UK, showing a strong mismatch between production and demand, which is reflected in waste, on the one hand, and exceeding emissions on the other, since resources are being spent on the production of goods that are not being absorbed by the market.

In more developed and developing countries where environmental policies are being taken more seriously, mainly due to the problems of pollution and climate change they are currently suffering, governments are imposing taxes, penalties and pollution quotas that oblige companies to adopt what is currently called green manufacturing. Indeed, more than 20 countries have already imposed carbon taxes to discourage industrial pollution and mitigate greenhouse gases emissions [3]. Some consumers who are more sensitive to the causes of the environment, tend to preferentially look for products that follow the principles of greener manufacturing [4]. However, for manufacturers to be able to lower their levels of greenhouse gas emissions associated with the manufacture of a given product, it is necessary that all their stakeholders agree with a plan, with a view also to reducing emissions in supply chains related to product manufacturing operations. The green supply chain is a structure focused on preserving the environment in its industrial and commercial activities, which includes suppliers, manufacturers, distributors, retailers, logistics operators, and customers, and in which all these members are committed to the same objective, assuming this shared responsibility [5]. The commitment to manufacture greener products can normally be divided into two distinct phases of the product's life: Green Product Manufacturing in the initial to mid-stage of product development, and Product Remanufacturing, in the final phase of the product's useful life. Green Product Manufacturing incorporates the principles of reducing the resources needed to produce the product and a production cycle that is significantly respectful of the environment, consuming less energy, producing less effluents and, in general, causing the least possible impact on the environment [6]. Environmental concerns should be respected right from the product design stage, thinking about the product's entire life cycle. The materials to be used, in addition to considering using the smallest amount possible, should also respect the environment, both in their extraction, processing or manufacture, as well as thinking about their reprocessing after the end of the product's life. The packaging should also be taken care of, as unnecessary volume means a greater number of trips for the same quantity of products, and the material used in the packaging, in addition to being used in the smallest amount possible, should preferably be recycled and therefore recyclable. Products that comply with these requirements, making due mention of the concerns that were taken in their design, manufacture, and packaging, will probably have a higher cost, but consumers will probably be predisposed to pay a higher price for these products, knowing that they deserved increased concerns with their environmental impact. However, manufacturing a product from scratch almost always consumes more resources than reconditioning existing and end-of-life products, namely, more energy, materials, and labor. In turn, the reconditioning of products involves a series of actions that have already been defined in advance, and which include the acquisition of these products in a used state, proceeding with their disassembly, inspection, cleaning, reconditioning and/or upgrade, reassembly, testing and repackaging [7]. Reconditioning should be a situation to be increasingly considered by producers and consumers alike. In fact, products can be designed modularly, bearing in mind possible evolutions of the same, or new requirements that may arise from consumers, adapting the product improving the features desired by customers, or replacing modules by others more evolved, maintaining the essence of the original product, but providing the new customers' requirements, according to the evolution of competitors, the market, and the consumers' desires. The definition of materials to be used in products is also crucial for increasing their longevity without leaving any traces of degradation due to its use, which will induce in the consumer a lesser desire to replace it. This way of acting allows companies to delineate reconditioning programs for their own products, thus inducing consumer loyalty. Simultaneously, its industrial activity will not be affected, as the labor used for remanufacturing is replacing the production of new products. Adequate planning of product reconditioning can allow the previously mentioned sequence of phases to be defined and programmed earlier, automating these processes, and making them as standardized as the manufacturing operations. Even in the case of remanufacturing operations, automation and robotics can play a critical role, through the creation of intelligent cells that can evaluate the state of the product and dismantling it in an organized way, to also allow an organized remanufacturing procedure, later. Kopacek [8] outlined some ideas to automate the dismantling of used products regarding their reconditioning, calling it as automated disassembling 4.0. The diagram corresponding to the principle drawn by Kopacek [8] can be seen in Fig. 4.1, comprising a set of equipment and automatic tools on the left, a set of logistics and manual tools on the right, and a set of sensors at the bottom. An operator can carry out a preliminary inspection of the used product, choosing whether the product is able to enter the disassembling cell or whether it first requires special attention. The sensors installed on the disassembly cell enable a more accurate diagnosis of the state of the product, even before starting the disassembly operations. The disassembly process should only interact with devices/components that need to be replaced or refurbished. The set of product support tools and jigs should be as flexible as possible, in order to make it possible to assemble as wide a set of products as possible, making the cell as flexible as possible. Ideally, these cells should be equipped with artificial vision systems for automatic detection of the type of product. The systems should also allow for operator intervention in circumstances where it becomes necessary, without it being necessary to stop the entire cell. This could imply the use of collaborative robots. This strategy will make it possible to reduce the return on investment and the occupation of space with different cells of this type.

The algorithm behind these operations should receive information from the sensors and, based on the principles established for remanufacturing, should perform the dismantling operations strictly understood as necessary, to restrict disassembly and remanufacturing operations to the essential minimum, depending on the characteristics that the company wants to provide to the consumers regarding the remanufactured product.

The advantages of refurbished products are obvious: (a) cost savings of around 50%, (b) a reduction in material consumption of around 70%, (c) a reduction in energy consumption of around 60% and (d) a reduction in greenhouse gas emissions of about 80% can be achieved [9]. Furthermore, refurbished products can provide a similar performance and quality as freshly manufactured products for the first time [10]. The price of refurbished products can be lower than the price presented by new products, encouraging customers to prefer refurbished products, without this negatively affecting the profit margins practised by companies, which can even benefit from customer loyalty and the perception that by purchasing this type of product they are contributing to a smaller ecological footprint, whether the company can successfully convey this perception to its customers [11]. Remanufacturing

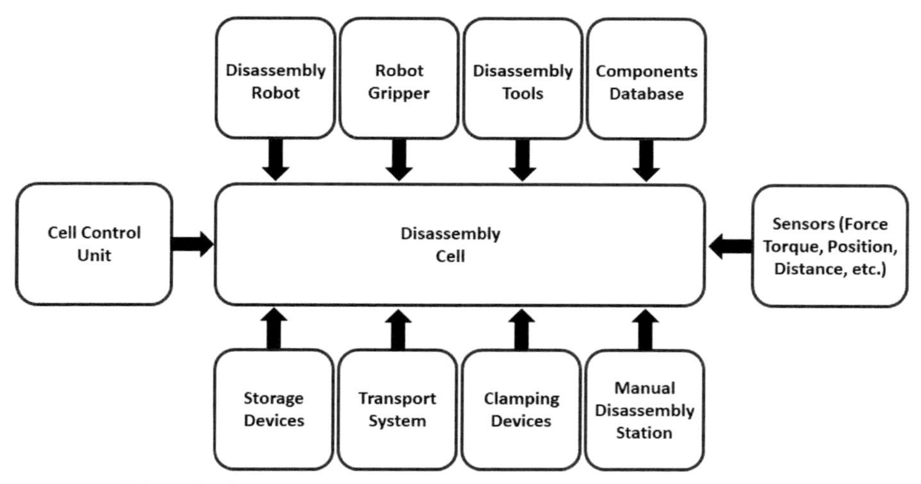

Figure 4.1. Schematic diagram of a cell designed for intelligent disassembly of used products aimed at remanufacturing.

may, however, present some difficulties in terms of market recognition, quality and satisfaction of requirements imposed by customers regarding other competing products, so presentation and price are determining factors in consumer persuasion.

As used products are not in the same state, this situation makes it difficult to adopt standardized procedures, which somewhat limits the remanufacturing work and brings some uncertainty to the remanufacturing operations. The logistical operations associated with remanufacturing are also a factor that can be negative but, if compared to the logistics of acquiring new materials, it can be equated, not representing a truly negative impact on the environment.

Remanufacturing still does not benefit from legislation, favoring its adoption in a more intensive way to the detriment of manufacturing fresh products. Thus, it becomes more attractive for companies to manufacture new products, spending more resources, than dedicating a substantial part of their efforts to preparation and remanufacturing operations. It is evident that Governments, from developed and developing countries, lack a fixed strategy in this matter, which promotes remanufacturing to the detriment of the manufacture of new products. Depending on the type of product, there is a clear margin of progression to evolve in this direction, promoting policies that benefit remanufacturing, which can be easily encouraged by a reduction in taxes applied to remanufactured products, among other ways of incentive that can be granted to this market, thus directly benefiting the environment.

It can therefore be seen that manufacturing new products following the concepts of green manufacturing or remanufacturing already used products will essentially depend on a few decisive factors: (a) the application of fees to companies based on their ecological footprint; (b) companies' capacity in terms of R&D; (c) companies' sensitivity to environmental issues; (d) environmental sensitivity of the companies' target market; (e) ability of the products to be easily updated/upgraded; (f) the level of market competitiveness in a given type of product. Only the combination of these factors allows for adequate weighting in the establishment of each company's strategy, opting for one or more solutions that satisfy the market in terms of quality and competitiveness, and that simultaneously help preserve the environment.

4.2 Greener production with reconfigurable manufacturing systems and robotics

Automation and robotics have been a pillar of increasing environmental sustainability in industrial terms. Automation and robotics have also been the mainstay of increasing industrial productivity, avoiding repetitive tasks for operators or operations that could pose danger to operators or professional illnesses, improving the production reliability and level of quality, reducing the space needed for product manufacturing and control, and integrate several processes in the same workstation, among many other factors. In fact, automation and robotics have benefited all three aspects of sustainability in the industry sector, although in different ways and with different degrees of intensity.

When talking about automation, we usually think of production or assembly lines dedicated to a product that is mass manufactured. Although automation started in that way, through the so-called fixed or hard automation [12], extremely important evolutions resulted from the most recent production needs, but which also opened the door to other production paradigms. Automation can either be implemented in an extremely complex production line, or in a single piece of equipment, called a stand-alone machine. However, automating the production of a product requires proper consideration of several factors, such as [12]:

- The type of product to be manufactured;
- The quantity and rate of the product demand (takt-time);
- Whether it is intended to automate the entire manufacturing process or just one or more parts;
- What is the level of qualification of the operators and technical staff, as well as their availability for training;
- What degree of reliability is intended for automated systems;
- What budget is available and the corresponding rate of return on investment.

Fixed or hard automation is costly and virtually inflexible. It is prepared to perform certain functions regarding a standardized product and, because the devices used in this type of automation are unprogrammable, or their programming capacity is reduced, each time it is necessary to make changes to the product, the devices need be changed or replaced, with the consequent disturbances to productivity and for commitments to deliver products to the market. Although this type of automation allows for adjustment of manufacturing parameters such as speed and cutting depth, it has limitations when it comes to accommodating other types of products or even the same product with significantly different dimensions (variants to the original product). As this type of equipment is quite expensive to design and build, this solution is only adopted when there is absolute certainty that the product will have good market acceptance and the quantities to be produced without any variation are quite high.

The design of fixed or hard automated equipment normally follows a modular logic, allowing one or more modules to be reused in equipment that may be designed in the future, when the current equipment has already fulfilled its mission and is out of service. In each module, an actuation principle (electrical, mechanical, pneumatic, hydraulic, etc.) is used to promote the work that these subsystems should perform. The accommodation and fixation of the product or semi-product during the operations triggered in each module are also critical, and condition flexibility since the jig for a given product may not allow another subproduct or component to be accommodated or remain conveniently fixed during the operations to be carried out. This type of automation is completely different to what can currently be seen in an automobile assembly line, where flexibility is very high. In fixed automation or hard-automation, flexibility, project complexity and global cost are factors that present strong limitations to its more generalized use, being very useful for standardized products and with high stability of features and form during a prolonged useful period. In view of permanent technological advances and changes in the market, where consumers began showing a greater enthusiasm for customizing products, the manufacture of products in large series became restricted to just a smaller range of products. In contrast, some products that were produced in large series, such as cars, now have a high number of variants. Fixed automation did not have the capacity to respond to these new market requirements, so it was necessary to evolve in terms of the type of automation to be used. In fact, it is not just the question of variability, but also the delivery time. Companies were faced with the need to deliver differentiated products, within the same family, but with very tight delivery times. This situation prevented the series from being more restricted and could be produced in sequence one from the other. In fact, it became necessary, within certain limits, to produce different products within a given family, but in a completely random sequence, defined by demand. In the first phase, the evolution of Programmable Logic Controllers (PLCs) brought great flexibility to automatic systems, allowing programming, and changing this programming in a much more accessible way. Automatic devices, mainly drives, began accepting different orders from PLCs, which immediately brought new windows of opportunity for automation, so that it ceased to be fixed and became flexible or programmable. Programming brought the possibility of working with different product shapes or sizes in the same automatic equipment, adjusting the programming, or even including image recognition devices capable of transmitting that information to the control system, which, through an appropriate algorithm, repositioned the systems and induced the necessary actions for the production or assembly of other product than that which was earlier being produced. The development of computers that provided greater capacity and command processing speed, as well as the creation of more user-friendly software, made it possible to speed up the process of making automation more flexible. Thus, process automation became available to a much larger number of users and came to be seen as a solution in cases where fixed automation could not be used because all devices could communicate coherently with each other. Indeed, communication protocols and interfaces capable of making interconnection possible were made available, increasing the interconnectivity between different brands and devices a lot.

Automation became popular, meeting the Lean principles. The Lean philosophy came to prove that mass-production is not the best approach for most of the productive cases existing in the market [13]. In fact, the diversity of products demanded by the market, which tends towards an increasingly intense customization, meant that the rigidity presented by production lines such as those of the Ford Model T no longer made sense, given the diversity of variants necessary to cover customers' requirements. Flexible or reconfigurable automation, also called soft automation [12], perfectly complies with what is established by Lean principles, significantly increasing process efficiency, drastically reducing the number of rejections due to the lack of quality, and fulfilling the need of only the produce what is ordered, avoiding the accumulation of products in stock. Additionally, the implementation of automation can also significantly enhance social sustainability, since, if the principles of ergonomics are complied with, there are no longer significant reasons for generating occupational diseases due to poor positioning at the workstation, tendinitis due to very prolonged cycles of repetitive efforts or movements, and even operations that could pose a risk to operators, such as handling parts that are still hot. Environmental sustainability can also be improved through automation. The greater reliability of automatic processes makes it possible to obtain a lower rate of defective parts, which, by itself, allows to state that the use of resources, both materials and energy, is carried out in a more efficient way. The precision allowed by automatic systems makes it possible to avoid some waste that would be natural due to greater uncertainty in human labor, which also enables material cost savings. The establishment of an optimized production rate allows a greater number of ready-to-market products to be obtained in the same period, which increases the efficiency of the process, optimizing the consumption of energy and other natural resources, such as water, for example. The reduction or non-existence of accidents at work, in addition to social sustainability, allows for greater work efficiency, allowing energy resources to be better used, which translates into resource savings, thus increasing environmental sustainability. A more accurate control of gaseous, liquid, and solid effluents, as well as adequate routing carried out by automatic systems, will also allow a more accurate notion of the pollution generated by each process, focusing attention on the most critical aspects, will enable these issues to be addressed more swiftly and effectively, preventing further harm to the environment.

Automation is essential to increase sustainability according to any of its pillars, and the evolution registered in the last two decades has allowed a greater adequacy to the needs of the market, expanding the possibility of using automation in sectors, in tasks and in size of lots where it was previously unthinkable that it could be equated and used. One of the main problems found in the industry when it is intended to alternate the manufacture of products is the time consumed in the reconfiguration of equipment, changing jigs, feeding of different materials or components, actuation, and sensing positioning, among other necessary tasks. None of this is needed on a former fully automated production or assembly line because it was designed with a single product in mind and was only used to produce that product. Any alteration of the product, if supported by the automated line, implied a lengthy change in tools, various adjustments, among others, of the line or equipment [14,15]. The development of computer systems and their better connection with the devices usually required by automation led to the development of software capable of creating a more user-friendly interface, which greatly facilitated the development of much more complex systems, but which are effectively in line with the real needs of most production systems. The integration of Artificial Intelligence in some algorithms has given even greater dynamics to this development, currently allowing fully automated and extremely flexible systems, where human intervention is usually unnecessary. With the modernization of automation, systems became easily reconfigurable, drastically reducing setup time, a problem previously identified by the Lean philosophy, and which has been partially circumvented through the SMED methodology [16-19]. This aspect of automation fostered to what is called Reconfigurable Manufacturing Systems (RMS), which are essentially based on three pillars: high quality, low cost, and high speed of response to changes. RMS, is made up of reconfigurable devices, tools, and machines, as well as material handling systems, previously prepared to allow easy and quick changing physical layouts to ensure sustainability. Below are three of the most informative definitions for RMS among the numerous definitions available:

- System designed to allow rapid changes in structure, as well as in hardware and software components, to quickly adjust production capacity and functionality within a given product family, effectively responding to sudden changes in market demand or legally imposed requirements;
- A production system that can be created by incorporating basic process modules, both hardware and software, which can be quickly and reliably reorganized or replaced, preserving product characteristics and quality;
- A manufacturing system based on modular architecture that uses hardware and software for quick, economical, and reliable reconfiguration or replacement of the production system.

RMS have shown a very high response capacity to the demands of modern industrial production. However, for this it is necessary to keep in mind some fundamental assumptions in the design of these systems, which include the following factors: (a) modularity; (b) scalability; (c) convertibility; (d) integrability; (e) customization, and (f) diagnostability. As this terminology is not always mastered by all those who are interested in automation and sustainability, each item mentioned above will be detailed below.

When considering modularity in the design phase consists of being able to divide a complex idea into much simpler subsystems, called modules, capable of partially manufacturing or assembling a product, which are not stand-alone by themselves, i.e., they only participate in the process, but not are a process in themselves. It may also be mentioned that these modules may be used in more than one context or equipment, being created to be independent, although they need consistency with the places where they are intended to be inserted. Designing modular systems implies knowing the processes very well and knowing how to assign certain functions to each subsystem, which depend on the preceding ones, giving rise to subsequent ones, without compromising the system globally. These subsystems should also be considered for possible future changes, thus minimizing fundamental changes, which could be restricted to a given module. This philosophy aligns with the principles of environmental sustainability, minimizing the use of resources. Connectivity to the preceding and succeeding subsystems is critical. It is also essential that the module be able to convey the product from the previous module to the next module, ensuring consistency in this movement. Additionally, it must also accommodate and properly secure the product during the operations that will be carried out in that module. Modularity allows for a much more effective approach to production and its management, as is the case when there is a need to modify the product, it may be necessary to replace one or another module, which can be done almost without any severe interruption of the production. In case the design is not modular, any modification that may be necessary could imply a much longer production stoppage and much higher costs, thus, it can be asserted that modularity fully satisfies at least two of the three pillars of sustainability: sustainability economic and environmental. Modularity can also help maintenance to increase economic and environmental sustainability, as repairing, reconditioning, or replacing a module is much more practical and economical than a complete line. The identification of the problem becomes faster, as the concept of each module is perfectly defined, its function is known, and it is easily understood which module is not fulfilling its function, allowing a faster intervention, using less resources. Meanwhile, studies have been conducted with the aim of improving environmental sustainability through energy savings in these RMS. In this manner, Massimi et al. [20] developed a mixed integer nonlinear mathematical model to optimize the energy consumption of an RMS, redefining two of its fundamental characteristics, namely, modularity and integrability. The results confirmed that modular machines have a positive impact on energy consumption when a set of auxiliary modules can perform different operations on the same equipment

Scalability is also an essential factor in modern production systems, given the variation in market demand. Scalability refers to the capacity of an RMS or production/assembly line to be designed and prepared efficiently to cope with a sudden increase in demand by swiftly adding or modifying resources or devices. For this scalability to be possible, the modules that make up each RMS should be properly studied in terms of lead-time and available space around them, so that the expansion can be planned and executed quickly. In order to optimize the efficiency of possible scalability solutions for RMS, Wang and Koren [21] developed an approach based on genetic algorithms to plan a production increase based on an existing production/assembly line, which allowed analyzing and selecting the most cost-effective solution to add resources to an RMS to attain-the capacity needed to meet demand. Scalability involves in-depth knowledge of the process and the exact perception of the places where the bottlenecks to the increase in production capacity are located, realizing whether acting in one or more modules will be enough to reach the determined goals. Tecnomatix Plant Simulation is an example of software that can simulate the workflow in RMS, enabling the study of various options and the scaling of the system where and when necessary.-Furthermore, the issue of available space and accessibility are fundamental, as the processes depend on the supply of materials and output semi-products or finished products that need to be properly disposed of to the next stage of production/assembly, or to the warehouse. The issue of equipment maintenance is also critical, both in terms of access and the diversity of critical spare parts. Proper consideration of environmental sustainability should begin from the design phase to ensure that the expansion of the RMS does not undermine the environmental benefits that were initially attained.

Due to market volatility, companies are often forced to change the features, performances, and look/shape of their products. Convertibility intends to translate the ease as an RMS system can be changed from one product to another, through the adaptation or change of equipment and/or tools, to meet the requirements necessary to satisfy the market. Several authors [22-24] have conducted studies to quantify the ease of conversion of RMS, typically validating their models with a case study, in order to quantify the ease of conversion. Basically, it can be said that convertibility largely depends on the possibility of changing the equipment used in the RMS, on how the way it is structured can be changed, and on the possibility of reconverting the layout.

Integrability concerns the ease with which new devices, systems or equipment can be integrated into the initial RMS, maintaining accuracy, product routing capability and inter-module communication capability. This property is extremely important when changing the product, or when there is a need to remedy any existing problem in the system. Integrability ensures that something can be changed in the RMS, keeping the focus on the quality and efficiency of the system as a whole, even after making the necessary changes. It becomes obvious that the devices, equipment, or systems to be introduced back into the RMS need to be properly studied and prepared to become coherent in terms of operation with the remaining modules that will remain in the RMS. This coherence will ensure that the reliability of the system as the whole will remain intact. The ease of integrating a new device, equipment or system will determine the degree of RMS integrability. The faster the adaptation to new RMS requirements and the lower the costs, the greater the RMS integrability. The integrability potential has already been studied by Farid [25] to analyze RMS and quantifying its integrability, allowing to understand whether in the future, if changes are necessary, what will be the workload necessary for the operations of integration of new devices, modules, equipment or systems. This indicator intends to measure the greater or lesser efficiency that the designers had in the first approach to the RMS, and what is the difficulty and workload that will be necessary when changes are necessary, which allows predicting the expected difficulties or facilities. It can also be said that this indicator measures the degree of freedom that a productive system presents, despite any necessary alteration. This integrability does not depend only on the mechanical systems, but on all hardware and software incorporated in the RMS, and the one that is intended to be integrated. It is in these factors that in most cases some incompatibilities could arise, which need to be previously and conveniently studied. The flexibility of communication between systems is fundamental in terms of hardware and software.

Customization is another extremely important property in RMS, and it can be defined as the capacity that the system should adapt to different variants of a product, within a previously stipulated family. The product is defined as a standard, around which several versions gravitate, depending on the selection criteria and options previously provided to the customer. In this way, the RMS and equipment and devices that constitute it should have the necessary flexibility to adapt to the manufacture of different variants of a base product [26]. The greater the flexibility of the production process, whether through the easy adaptation of hardware or software, the greater the degrees of freedom to extend the product family and meet the customer's wishes, within the range of options considered. This production system architecture is also an environmentally sustainable solution, as it minimizes the necessary resources, allowing the manufacture of a range of products in the same space and production system, with minimal adjustments (tools). The design of customizable systems has also drawn the attention of researchers. Pugliese et al. [27] developed a methodology that helps define the best configuration of production systems in order to increase the degree of customization of an RMS. The methodology allows structuring the thinking of those who design the systems, establishing an architecture that favors the flexibility of the system to adapt to a greater variety of product variants, requiring less adaptation effort. The index that is determined by the developed model allows estimating the impact that the flexibility and reconfigurability of an RMS can have on its own performance. The same model can also be applied to specific tools, molds, stamping dyes and jigs [28].

Diagnosability is another of the pillars on which RMS is based. The design of these systems tries to inventory all the possible sources of uncertainty, to reduce as much as possible the risk of quality problems and lack of reproducibility, increasing the reliability of the production systems. In terms of environmental sustainability, this aspect is extremely important, as it is directly linked to the waste of resources and even the possible generation of scrap, which, in some cases, is not even recyclable (some thermosetting polymers, etc.). Thus, to predict possible deviations in production systems involving multiple variables, Maler-Speredelozzi et al. [29] applied the Design for Diagnosis method. With the developed model, they intended to identify the critical design conditions that allowed obtaining the highest possible reliability in the developed production system. The Design for Diagnosis methodology is based on optimizing the production efficiency and performance of automatic production systems in different industrial scenarios. The same authors concluded that diagnosability directly depends on the number of equipment and devices integrated in the production system, being worse the greater the number of devices and equipment considered. Diagnosability is an extremely important factor in defining the level of quality that can be expected from a complex automatic production system, which is why it should assume special relevance in system analysis. Additionally, it is also extremely important for environmental sustainability, as lower diagnosability implies lower quality, greater waste of resources and more scrap generated, or greater recovery effort (energy and other resources) to be put into defective products.

RMS require quite intense prior planning, given their complexity and regarding that they are a multi-objective problem, given that they must meet lead-time, takt-time, quality, product production cost, flexibility, and respect for the environment. The complexity is a result of (a) all the requirements at the base of the project, (b) the limitations imposed by the equipment and devices used in the design of the RMS, (c) the establishment of the necessary degrees of freedom able to allow its reconfiguration, modification, or expansion in the future, (d) the respect by the imposed quality criteria, (e) the fulfillment of cost limits, and (f) the multiplicity of modules that can be aggregated. Each module will have its own specificities and limitations, but it needs to be perfectly aligned with the rest, so that the whole achieves the intended results. Several approaches have been developed to the problems of configuration and reconfiguration of these productive systems, which have gone through the most diverse methodologies. Yamada [30] used the particle sworm optimization methodology to design, position and establish the necessary manufacturing/assembly time in each module, planning and establishing time forecasts for the dynamics of reconfiguration of productive systems. Other researchers have carried out much more complex approaches, comparing the results of different methodologies in planning the process to be carried out by RMS, using methodologies such as Simulated Annealing, Archived Multi-Objective Simulated Annealing and Generic Algorithms, analyzing which methodology offers better solutions in the planning of reconfigurable production systems [31]. However, the most used methodology to optimize the design and operation of RMS is the non-dominated sorting genetic algorithm II (NSGA-II) [32].

The advent of robotization has expanded the horizons initially provided by automation. The incorporation of stepper and servo-controlled motors has enabled more precise and straightforward movement control through programming. The strong increase in sensing has, on the other hand, allowed an almost real-time verification of the movements of the automatic systems, which has resulted in a much more effective and reliable control. Robotics, when aided by flexible automatic systems, makes production systems much simpler to implement and allows for much more accurate control of any process.

Despite the efficiency already provided by these systems, which was further increased in the last two decades by the increase in sensing, making cooperation between the work carried out by robots and by humans perfectly reconcilable, it is still necessary to go further, placing all this technology to the services of the environment. If Lean practices are mainly concerned with economic efficiency and sustainability, there has also been a new aspect associated with Lean practices, called Lean-Green, which promotes environmental sustainability together with economic sustainability. This translates, in practice, into making processes more efficient and, simultaneously, saving resources and avoiding the pollution generation.

Robotics, largely supported by automation and information technology, have been at the base of what is called Smart Manufacturing (or Industry 4.0), which, due to its current importance for sustainability, will be dealt with in a specific chapter.

4.3 Sustainable manufacturing and lean green

Given the three pillars that support sustainability, economic, environmental, and social, it is not easy to find a sufficiently comprehensive definition for Sustainable Manufacturing, as the efforts made could be focused only on a pillar, on two, or even on all of them. Furthermore, even if the focus is on only one of the pillars, it is not clear whether there is a positive or negative influence on each of the remaining pillars. It is true, however, that most of the efforts relating to sustainability, in the beginning, focused essentially on the economic aspect, trying to reduce costs. However, this cost reduction can both benefit and harm environmental sustainability.

In fact, the substitution of some materials for other more economical ones can increase economic sustainability, but cause problems for the environment. However, the opposite can also occur. Efforts to reduce electricity consumption simultaneously improve economic and environmental sustainability. Therefore, defining Sustainable Manufacturing is an almost impossible task, due to different interactions. However, it is common sense that Sustainable Manufacturing should incorporate improvements according to the abovementioned three pillars, improving working conditions, reducing the environmental impact, and allowing companies to produce more competitive products.

Economic sustainability can easily be seen as a single objective, as this has been the focus of numerous actions that were duly recorded in the past, through numerous case studies reported, in which the focus of the study was essentially to make it more competitive. It is not always perfectly clear whether the result of the savings achieved in these studies accrued to companies or consumers. However, it would be good if the profits reverted in favor of the companies, this money was invested in social and environmental concerns, which would no longer create wealth for a restricted group of shareholders and would start to be useful to a much wider group of citizens. This idea may seem somewhat utopian, but it cannot be ignored and deserves to be mentioned, as it can inspire the thinking and management approach of some companies.

Mittal and Sangwan [33] analyzed and ranked 13 drivers leading to the adoption of Sustainable Manufacturing practices using the Extended Fuzzy TOPSIS methodology, concluding that four of these drivers exerted a greater influence on the adoption of these practices. These authors concluded that the most influential drivers are competitiveness between companies, motivations induced by current legislation, organizational resources technology. The ideas presented by Kumar et al. [34] argued that implementing sustainable manufacturing practices is dependent on the size of the sector where the company operates, the type of product produced, and the development capacity of potential suppliers of solutions for greener manufacturing, making it a controversial and difficult idea to sustain. However, these ideas are questionable since the size of the sector has little impact on a company's ability to develop environmentally friendly solutions. Ultimately, it is the responsibility of the company's top management to establish strategies, make investments in technology and development, and facilitate the pursuit of more ambitious goals in terms of environmentally friendly practices.–However, it cannot be categorically stated that the product conditions the adoption of Sustainable Manufacturing practices. Moreover, it is the responsibility of the company to develop alternative products, using different materials and/or manufacturing processes, which present a smaller ecological footprint, benefiting the consumer and the environment. Technology can be limiting for some developments, but research around the limiting factors for greener production are constantly producing results, which translate into new solutions that, if focused on the environment, will certainly lead to the implementation of more sustainable production practices.

Garbie [35] dissected the requirements that industrial companies should present to be considered sustainable, raising a concept of sustainable economic-social-environmental development and systematic practices related to the sustainable creation of value, considering the economic, social and environmental aspects, through the development of products, processes (engineering), and ergonomics, health and safety at work components, establishing appropriate policies. Aljuneidi and Bulgak [36] developed a theoretical model to design new production systems that would allow the connection between the company's management assumptions, the implementation of environmental sustainability practices, and the efficiency of these environmental practices in the context of reconfigurable hybrid manufacturing systems, simultaneously considering production systems and remanufacturing. The authors stated that organizational culture and internal procedures largely influence design freedom. In many situations, taking care of the environmental aspect implies sacrificing the cost of the product or the process. In this matter, the company's culture and guidelines become decisive, allowing or not its product development department to take more daring options that allow safeguarding the environment, to the detriment of having as its main focus the cost and competitiveness in terms of price and profit yield. In fact, if companies conveniently explain their mission and position in the development of their products, describing how their practices meet the needs of preserving the environment, these will be arguments that will certainly sensitize a large part of consumers, as they are already feeling and suffering the consequences of the carelessness and passivity that guided the concerns of companies during the second half of the 20th century.

The design stage plays a crucial role in the entire life cycle of a production system. There are certain factors that should be considered when defining the production system, such as: (a) the useful life of the system should be as long as possible, although this may depend on the interest of the product on the market; (b) the devices, systems or equipment to be used should be as durable as possible to be able to be integrated into future systems to be designed in the future, after the useful life cycle of the current production system; (c) the amount of effluent emissions should be as low as possible; and (d) energy consumption should be as low as possible. Therefore, the selection of devices, tools, and equipment to be integrated into the production system should be as smart as possible, allowing environmental values to be preserved. Usually, there is a maximum budget established for each project, but this should be conveniently negotiated and used, allowing the most environmentally friendly technologies to be selected, to the detriment of others that could favor economic competitiveness, but does not allow for arguments around the environmental defense.

Energy efficiency has been the subject of many studies, given that it is an economic and environmental concern at the same time. However, some difficulties are hard to overcome when trying to optimize energy efficiency. These difficulties are described and explained below:

- During the design phase, a maximum limit can be established for the energy consumption of a given RMS. However, this system is made up of several modules. The distribution of consumption by the different modules is not always consensual, considering the specific needs of each task that should be taken into consideration, as well as the intended way in which the task will be carried out. Creativity in terms of Engineering and a strict discipline in fulfilling the initial objectives should be maintained, as it is common for the design of different modules to be allocated to different project teams, which could lead to a lower commitment to the initially established objectives;

- Global consumption goals have implications for selecting which modules to use, and which modules to use have implications for global consumption. It is essential that the focus remains unwavering in terms of reducing consumption as much as possible, avoiding all tasks that may seem unnecessary, after a critical reflection on the initial ideas for the project;
- Any reconfiguration will have to consider the initial requirements of minimum energy consumption, and should also prevent energy consumption peaks from exceeding values that could jeopardize the integrity of the electrical installation and the objectives of reducing energy consumption;
- Given that energy consumption targets are dynamic over time and should reflect economic and environmental concerns that become worse over time, reactive models should be proposed that make it possible to automatically adjust the RMS as a function of new energy impositions/limitations that could be set by regulation or by guidelines issued by top management.

Given that the useful life of automated production systems is usually several years, and that during this useful life the regulations can be updated, as well as the product features, the RMS need special care during their conception, trying to facilitate the upgrades to which they will almost necessarily have to undergo during their useful life. Designing defensively, exceeding the requirements that are in place at the time of design can be a great strategy, minimizing time and effort to upgrade in the future. Modularization assumes great importance here again, as technological development itself could dictate that the life of the RMS can still be extended only with certain technological updates, sufficient to comply with current regulations and keep the product competitive in the market. This strategy will allow the principles of environmental sustainability to be met, as the use of new resources will be avoided in a shorter time.

Lean practices have been associated with a green posture of companies, which can bring dividends in terms of the market, increasing the visibility of companies in an increasingly demanding market, given the perception of the environmental impact that production systems can have. Teixeira et al. [37-39] studied the relationship between the Lean philosophy and green practices, concluding that the increase in the maturity of the implementation of Lean tools progressively contributes to the strengthening of environmental sustainability, through a greater focus on Green practices, emphasizing the Lean-Green connection. Effectively, deepening of green practices can significantly strengthen the stakeholders' engagement and can set new goals to meet increasingly demanding environmental standards, thereby enhancing their sustainable performance and improving their reputation in the market. These authors identified a strong relationship between the Lean philosophy and green practices, both equally with a strong influence in the pursuit of what is defined as the Triple Bottom Line, i.e., the three pillars of sustainability: economic, environmental, and social, inducing a strong competitive advantage in companies able to conveniently associate Lean with Green. This study was confined to Portugal, but an idea was clearly highlighted: a greater competitive advantage can be obtained when the two management techniques are applied jointly, rather than exploring each one separately, and this idea is also reported in other works [40,41]. In a study presented in 2021, Teixeira et al. [38] found just 23 studies about the integration of Green in Lean philosophy, which is very little given the relevance of the topic for the future of the industry. Zhan et al. [42] studied the integration of these concepts in 172 Chinese companies concluding that it leads both to an economic and environmental improvement, namely through a reduction in the consumption of hazardous materials, reduction of solid waste generation, air emissions, wastewater, as well as environmental accidents. However, the number of studies relating Lean-Green with sustainability is still scarce, showing that a gap in research has been identified, which can still be largely explored.

The automotive industry is the one that has promoted more studies on this subject. Despite being an extremely competitive industry, its size and economic importance, as well as fierce competition for the market, has led to the fact that studies have multiplied in this sector of activity. However, product development is also extremely important in this sector, as it is common sense that the mobility associated with internal combustion engines is a strong generator of greenhouse gases, and a weight reduction in everything that is related to mobility, decreases the level of greenhouse gas emissions. Thus, the mobility sector in general, and the automotive industry, in particular, cannot underestimate the application of Lean and Green principles in their activities, to keep them competitive.

It is also possible to observe that, within the Triple Bottom Line, the economic and environmental pillars tend to be jointly increased when Lean is supported by Lean practices, but it is more difficult for companies to associate the social aspect. In fact, considering a study developed by Morgado et al. [43] taking the ISO 45001:2018 standard as a support, that while around 98% of companies acknowledge the importance of having plans for hygiene, safety, and health at work, only 75% of the companies in the sample considered had actual plans for monitoring risks, preventing accidents, and other related measures. This aspect may not have been correctedly portrayed in these studies, because the sample considered in the study excludes a significant number of highly automated industries. Automation and robotics, if properly applied, also have a very strong impact on the social sustainability component [44-48], in association with ergonomic and work safety studies that should be performed when designing highly automated production systems.

References

[1] World Economic Forum. (2022). Reducing the carbon footprint of the manufacturing industry through data sharing. URL: https://www.weforum.org/impact/carbon-footprint-manufacturing-industry/, (Accessed on December 31st, 2022).

[2] World Heath Organization. (2022). Air pollution. URL: https://www.who.int/health-topics/air-pollution#tab=tab_2, (Accessed on January 1st, 2023).

[3] Luo, R., Zhou, L., Song, Y. and Fan, T. (2022). Evaluating the impact of carbon tax policy on manufacturing and remanufacturing decisions in a closed-loop supply chain. International Journal of Production Economics, 245: 108408. doi: 10.1016/j.ijpe.2022.108408.

[4] Zhang, X. and Yousaf, H.A.U. (2020). Green supply chain coordination considering government intervention, green investment, and customer green preferences in the petroleum industry. Journal of Cleaner Production, 246: 118984, doi: 10.1016/j.jclepro.2019.118984.

[5] Wang, Y., Xu, X. and Zhu, Q. (2021). Carbon emission reduction decisions of supply chain members under cap-and-trade regulations: a differential game analysis. Computers in Industrial Engineering, 162: 107711. doi: 10.1016/j.cie.2021.107711.

[6] Fraccascia, L., Giannoccaro, I. and Albino, V. (2018). Green product development: what does the country product space imply? Journal of Cleaner Production, 170: 1076–1088. doi: 10.1016/j.jclepro.2017.09.190.

[7] Teixeira, E. L. S., Tjahjono, B., Beltran, M. and Julião, J. 2022. Demystifying the digital transition of remanufacturing: a systematic review of literature. Computers in Industry, 134: 103567 doi: 10.1016/j.compind.2021.103567.

[8] Kopacek, P. (2019). Trends in production automation. IFAC PapersOnLine, 52(25): 509–512. doi: 10.1016/j.ifacol.2019.12.595.

[9] Okorie, O., Obi, M., Russell, J., Charnley, F. and Salonitis, K. (2021). A triple bottom line examination of product cannibalisation and remanufacturing: A review and research agenda. Sustainable Production and Consumption, 27: 958–974. doi: 10.1016/j.spc.2021.02.013.

[10] Zhu, X., Wang, M., Pei, J. and Pardalos, P. M. (2020). Investigating remanufacturing competition with yield uncertainty on market share, profit, and consumer surplus. International Transactions in Operational Research, 27(5): 2584–2615. doi: 10.1111/itor.12712.

[11] Ilgin, M. A. and Gupta, S. M. (2012). Remanufacturing Modeling and Analysis. CRC Press, Boca Raton, FL, USA. ISBN: 978-0429104831. doi: 10.1201/b11778.

[12] Kalpakjian, S. and Schmid, S. R. (2001). Manufacturing Engineering and Technology, 4th Ed., Prentice Hall, Hoboken, NJ, USA. ISBN: 0-13-017440-8.

[13] Silva, F. J. G. and Ferreira, L. P. (2019). Lean Manufacturing: Implementation, Opportunities and Challenges. Nova Science Publishers, NY, USA. ISBN: 978-1-53615-725-3.

[14] Rosa, C., Silva, F. J. G., Ferreira, L. P. and Campilho, R. (2017). SMED methodology: The reduction of setup times for Steel Wire-Rope assembly lines in the automotive industry. Procedia Manufacturing, 13: 1034–1042. doi: 10.1016/j.promfg.2017.09.110.

[15] Sousa, E., Silva, F. J. G., Ferreira, L. P., Pereira, M. T., Gouveia, R. and Silva, R. P. (2018). Applying SMED methodology in cork stoppers production. Procedia Manufacturing, 17: 611–622. doi: 10.1016/j.promfg.2018.10.103.

[16] Martins, M., Godina, R., Pimentel, C., Silva, F. J. G. and Matias, J. C. O. (2019). A practical study of the application of SMED to electron-beam machining in automotive industry. Procedia Manufacturing, 17: 647–654. doi: 10.1016/j.promfg.2018.10.113.

[17] Vieira, A. M., Silva, F. J. G., Campilho, R. D. S. G., Ferreira, L. P., Sá, J. C. and Pereira, T. (2020). SMED methodology applied to the deep drawing process in the automotive industry. Procedia Manufacturing, 51: 1416–1422. doi: 10.1016/j.promfg.2020.10.197.

[18] Monteiro, C., Ferreira, L. P., Fernandes, N. O., Sá, J. C., Ribeiro, M. T. and Silva, F. J. G. (2019). Improving the machining process of the metalworking industry using the lean tool SMED. Procedia Manufacturing, 41: 555–562. doi: 10.1016/j.promfg.2019.09.043.

[19] Santos, V., Sousa, V. F. C., Silva, F. J. G., Matias, J. C. O., Costa, R. D., Pinto, A. G. and Campilho, D.G.S.G. (2022). Applying the SMED methodology to tire calibration procedures. Systems, 10: 239. https://doi: 10.3390/systems10060239.

[20] Massimi, E., Khezri, A., Benderbal, H. H. and Benyoucef, L. (2020). A heuristic-based non-linear mixed integer approach for optimizing modularity and integrability in a sustainable reconfigurable manufacturing environment. The International Journal of Advanced Manufacturing Technology, 108(7): 1997–2020. doi: 10.1007/s00170-020-05366-y.

[21] Wang, W. and Koren, Y. (2012). Scalability planning for reconfigurable manufacturing systems. Journal of Manufacturing Systems, 31(2): 83–91. doi: 10.1016/j.jmsy.2011.11.001.

[22] Maler-Speredelozzi, V., Koren, Y. and Hu, S. J. (2003). Convertibility measures for manufacturing systems. CIRP Annals, 52(1): 367–370. doi: 10.1016/S0007-8506(07)60603-9.

[23] Lafou, M., Mathieu, L., Pois, S. and Alochet, M. (2014). Convertibility indicator for manual mixed-model assembly lines. Procedia CIRP, 17: 314–319. doi: 10.1016/j.procir.2014.03.113.

[24] Chinnathai, M. K., Alkan, B. and Harrison, R. (2017). Convertibility evaluation of automated assembly system designs for high variety production. Procedia CIRP, 60: 74–79. doi: 10.1016/j.procir.2017.01.005.

[25] Farid, A. M. (2017). Measures of reconfigurability and its key characteristics in intelligent manufacturing systems. Journal of Intelligent Manufacturing, 28: 353–369. doi: 10.1007/s10845-014-0983-7.
[26] Wang, G. X., Huang, S. H., Yan, Y. and Du, J. J. (2017). Reconfiguration schemes evaluation based on preference ranking of key characteristics of reconfigurable manufacturing systems. The International Journal of Advanced Manufacturing Technology, 89(5): 2231–2249. doi: 10.1007/s00170-016-9243-7.
[27] Pugliese, V., Mesa, J. and Maury, H. (2017). Development of a design methodology for reconfigurable injection molds. The International Journal of Advanced Manufacturing Technology, 90(1): 153–166. doi: 10.1007/s00170-016-9348-z.
[28] Nishith Mohan, F., Rishi, G. and sharma, S. (2013). Flexibility measurement criteria with respect to reconfigurable system properties. International Journal of Engineering Research and Application, 3(5): 1711–1716. doi: N/A.
[29] Maler-Speredelozzi, V., Koren, Y. and Hu, S. J. (1997). Stream-of-variation theory for automotive body assembly, CIRP Annals, 46(1): 1–6. doi: 10.1016/S0007-8506(07)60603-9.
[30] Yamada, Y. (2006). Dynamic reconfiguration of reconfigurable manufacturing systems using particle swarm optimization. Proceedings 2006 IEEE International Conference on Robotics and Automation (ICRA 2006), Orlando, FL, USA, 2006, 1444–1449, doi: 10.1109/ROBOT.2006.1641912.
[31] Touzout, F. A., and Benyoucef, L. (2019). Multi-objective sustainable process plan generation in a reconfigurable manufacturing environment: Exact and adapted evolutionary approaches. International Journal of Production Research, 57(8): 2531–2547. doi: 10.1080/00207543.2018.1522006.
[32] Brahimi, N., Dolgui, A., Gurevsky, E., and Yelles-Chaouche, A. R. (2019). A literature review of optimization problems for reconfigurable manufacturing systems. IFAC-PapersOnLine, 52(13): 433–438. doi: 10.1016/j.ifacol.2019.11.097.
[33] Mittal, V. K. and Sangwan, K. S. (2015). Ranking of drivers for green manufacturing implementation using fuzzy technique for order of preference by similarity to ideal solution method. Journal of Multi-Criteria Decision Analysis, 22(1-2): 119–130. doi: 10.1002/mcda.1527.
[34] Kumar, M., Srai, J. and Minakata, Y. (2017). Industrial system dynamics for environmental sustainability: A case study on the UK Medical Technology sector. Industrial system dynamics for environmental sustainability: A case study on the UK medical technology sector, 31 (1/2/3), 100-132. doi: 10.1504/IJMTM.2017.10002932.
[35] Garbie, I. H. (2014). An analytical technique to model and assess sustainable development index in manufacturing enterprises. International Journal of Production Research, 52(16): 4876–4915. doi: 10.1080/00207543.2014.893066.
[36] Aljuneidi, T. and Bulgak, A. A. (2016). A mathematical model for designing reconfigurable cellular hybrid manufacturing-remanufacturing systems. The International Journal of Advanced Manufacturing Technology, 87(5): 1585–1596. doi: 10.1007/s00170-016-9141-z.
[37] Teixeira, P., Coelho, A., Fontoura, P., Sá, J. C., Silva, F. J. G., Santos, G. and Ferreira, L. P. (2022). Combining lean and green practices to achieve a superior performance: The contribution for a sustainable development and competitiveness—An empirical study on the Portuguese context. Corporate Social Responsibility and Environmental Management, 29(4): 887–903. doi: 10.1002/csr.2242.
[38] Teixeira, P., Sá, J. C., Silva, F. J. G., Ferreira, L. P., Santos, G. and Fontoura, P. (2021). Connecting lean and green with sustainability towards a conceptual model. Journal of Cleaner Production, 322: 1290047. doi: 10.1016/j.jclepro.2021.129047.
[39] Teixeira, P., Sá, J. C., Silva, F. J. G., Santos, G., Fontoura, P. and Coelho, A. (2021). Lean contribution to the companies' sustainability. IFIP International Federation for Information Processing 2021, Springer Nature Switzerland AG, D. J. Powell et al. (Eds.): ELEC 2021, IFIP AICT 610, 400–408. doi: 10.1007/978-3-030-92934-3_42.
[40] Silva, S., Sá, J. C., Silva, F. J. G., Ferreira, L. P. and Santos, G. (2020). Lean green–the importance of integrating environment into lean philosophy–a case study. *In*: Rossi, M., Rossini, M. and Terzi, S. (eds.). Proceedings of the 6th European Lean Educator Conference. ELEC 2019. Lecture Notes in Networks and Systems, 122: 211–219. Springer, Cham, Switzerland. doi: 10.1007/978-3-030-41429-0_21.

[41] Dües, C. M., Tan, K. H. and Lim, M. (2013). Green as the new lean: How to use Lean practices as a catalyst to greening your supply chain. Journal of Cleaner Production, 40: 93–100. doi: 10.1016/j.jclepro.2011.12.023

[42] Zhan, Y., Tan, K. H., Ji, G. and Tseng, M.-L. (2018). Sustainable Chinese manufacturing competitiveness in the 21st century: Green and lean practices, pressure and performance. International Journal of Computer-Integrated Manufacturing, 31: 523–536. doi: 10.1080/0951192X.2016.1268721.

[43] Morgado, L., Silva, F. J. G. and Fonseca, L. M. (2019). Mapping occupational health and safety management systems in portugal: Outlook for ISO 45001:2018 adoption. Procedia Manufacturing, 38: 755–764. doi: 10.1016/j.promfg.2020.01.103.

[44] Costa, R. J. S., Silva, F. J. G. and Campilho, R. D. S. G. (2019). A novel concept of agile assembly machine for sets applied in the auto-motive industry. International Journal of Advanced Manufacturing Technology, 91: 4043–4054. doi: 10.1007/s00170-017-0109-4.

[45] Costa, M. J. R., Gouveia, R. M., Silva, F. J. G. and Campilho, R. D. S. G. (2018). How to solve quality problems by advanced fully-automated manufacturing systems. International Journal of Advanced Manufacturing Technology, 94: 3041–3063. doi: 10.1007/s00170-017-0158-8.

[46] Silva, F. J. G., Swertvaegher, G., Campilho, R. D. S. G., Ferreira, L. P. and Sá, J.C. (2020). Robotized solution for handling complex automotive parts in inspection and packing. Procedia Manuf. 2020, 51: 156–163. doi: 10.1016/j.promfg.2020.10.023.

[47] Silva, F. J. G., Soares, M. R., Ferreira, L. P., Alves, A. C., Brito, M., Campilho, R. D. S. G. and Sousa, V. F. C. (2021). A novel automated system for the handling of car SeatWires on plastic over-injection molding machines. Machines, 9: 141. doi: 10.3390/machines9080141.

[48] Araújo, L. M. B., Silva, F. J. G., Campilho, R. D. S. G. and Matos, J. A. (2017). A novel dynamic holding system for thin metal plate shearing machines. Robotics and Computer-Integrated Manufacturing, 44: 242–252. doi: 10.1016/j.rcim.2016.06.006.

Chapter 5

Smart Manufacturing as a Way for Greener Processes

5.1 A general perspective

In recent years, smart manufacturing has been associated with intelligent manufacturing systems, managed by powerful information systems, which interconnect the actions of different equipment and monitor the evolution of operations, allowing a faster and more fluid manufacturing sequence, while also collecting information on the evolution of the product in this manufacturing sequence. This makes it possible to have an exact notion of the status of each production order, detect or even anticipate equipment problems, and have real-time information on necessary maintenance actions. In summary, there is a perfect symbiosis between the equipment, production scheduling, and information systems, which, after a careful study of interconnecting the systems and programming allows for almost perfect control of production lines, even when endowed with some complexity. In this chapter, the concept of smart manufacturing is much broader, essentially aiming to highlight automated manufacturing processes that use technology to reduce their environmental impact.

Within this broader sense, special emphasis will also be given to processes in the chemical industry, due to the environmental impact that traditional processes normally imply, as well as to additive manufacturing, because it is a manufacturing process that has caused a disruption in the metalworking industry, mainly in terms of how complex parts can be manufactured in a much easier way, without the existence of molds or tools, with completely different logistics, and starting to allow local production, to the detriment of centralized production systems and, therefore, sometimes extremely distant [1,2]. In the case of the chemical industry, known for several extremely serious environmental disasters and for the health and environmental problems it can create, some evolutions in processes will be discussed here that started to take the environment more into account and, with that, the health of the workers involved in the processes. In the case of additive manufacturing, the evolution is tremendous and everything that is said at this moment may be outdated soon, but it is essentially intended to draw attention to how this manufacturing process, not being exempt from criticism in terms of the environment and health of the workers involved, allows for a much tighter control of environmental conditions, and may even be a promoter of the environment, in terms of the use of recycled materials.

A brief review of other processes understood as important within the scope of this book is also carried out. However, the main focus of this chapter is the chemical industry and additive manufacturing based on very recent works.

The work developed by Lyu et al. [3] highlighting the main virtues of adopting smart manufacturing, essentially focusing on the Chinese industry. These authors indicated that the adoption of smart manufacturing promotes innovation in terms of the search for more environmentally friendly technologies, noting also that this will also translate into an increase in the companies' operating profit. Although the Chinese case may have slightly different contours from other highly industrialized countries, the study concludes that, in addition to the market differentiating factors that companies oriented toward smart manufacturing will be able to take advantage of, there may also be government support in this regard through programs specifically oriented to encourage the development of companies with much higher technological indices. Moreover, companies will also be able to obtain more attractive commercial yields due to government support, and a more competitive production system. However, this study points to a significant difference in behavior and results within the Chinese territory, that is, the implementation of smart manufacturing pilot projects in companies in the eastern region presents better results than those in the central and western regions. This conclusion allows us to understand that the character of managers and workers, their initial training, and the ability to learn new concepts, among others, can be key factors for the successful implementation of smart manufacturing, especially when it is intended to achieve good indicators environmental and economic. Another conclusion of this study also reveals another important factor: even the gender of the main manager can interfere with success, with companies managed by female people presenting better results. Larger research and development teams, whose proportion in relation to the overall number of employees in the companies is more favorable, also translates into a greater success rate. These conclusions suggest a possible high heterogeneity in terms of success in the implementation of smart manufacturing, with a focus on more environmentally friendly production. Therefore, since the results of this study cannot be directly extrapolated, it allows us to have an idea that each market should be studied per se, to understand whether smart manufacturing has an effective positive influence on the companies' environmental indicators.

As in other works that have studied the evolution of smart manufacturing in China, Abudureheman et al. [4] noted that there is some heterogeneity in adherence to this concept of production in different regions of China. However, the best conclusion to be drawn from this work is that smart manufacturing has contributed to the reduction of carbon emissions. According to these authors, this is essentially due to two factors: (a) the technological advancement of new manufacturing processes, due to their greater efficiency; (b) scale effects. This work does not refer to the logistical factor as preponderant, but it should not be ignored, since the integration of processes leads to an effective reduction of logistical operations within each company, which translates into significant gains in time, elimination or reduction of waiting times between operations (with consequent gains in lead-time), and transport of semi-products between the different equipment in the production sector, a situation that is now managed automatically by electrically or pneumatically driven conveyors, with a clear reduction in energy needed, and with increased efficiency.

Abudureheman et al. [4] also identified some important indicators that will allow one to understand whether a region, country, or company is properly prepared to evolve toward a smart manufacturing concept. These indicators are as follows: (a) infrastructure construction; (b) talent input; (c) research and development investment intensity; (d) artificial intelligence technology; (e) big data and data analytics technology; (f) cloud computing technology; (g) internet of things technology. However, the last four indicators may be included in the first one. Effectively, the creation of infrastructure is basic for the adoption of smart manufacturing, as the last four factors are essentially related to information systems, which largely depend on infrastructure, although not exclusively. Due to the high volume of data generated by smart manufacturing and the strong interconnection that they need to have, the support information systems need to support and know how to manage this information, since the objective is also to minimize the need for control by humans. Thus, the necessary data analysis systems should be created so that only the most relevant information for production management is issued. Talent is something perfectly implicit in smart manufacturing. In fact, these systems need to be widely studied and discussed to obtain the best solutions. For this, technicians need to have high technical capability to understand and minimize production operations, link them in the best way, and select production and interface devices for feeding, handling, monitoring and control, and packaging, ensuring a smooth flow of the product as 'lean' as possible. Minimizing the generation of defects is also a concern for technicians. Therefore, for the installation of a smart manufacturing system, skills in various fields of knowledge are required, namely in materials, manufacturing processes, automation, programming metrology, quality, and management. These valences require very well-prepared and highly talented professionals. For smart manufacturing to proliferate in the global industrial sector, it is necessary that technical training, mainly at the university level, be highly focused on capturing and providing resources with these valences to the industry. To this end, the dissemination of smart manufacturing among secondary schools is essential, inducing potential future professionals to want to follow this path, realizing that the market will need them and that they will have a very noble role in terms of increasing productivity and essentially improving the environment in the future.

Smart manufacturing has another considerable advantage: the close monitoring of processes makes it possible to detect non-conformities at any stage of production, which can lead to: (a) stopping the production process immediately, preventing defective components from being produced, and avoiding rework or scrap metal, with all resource consumption being wasteful; (b) easy recovery of the semi-product through rework. These possibilities make it possible to save energy that would be spent producing components that would not conform, and to save material resources that would be wasteful, as they would have to be reworked, recycled, incinerated, or deposited in landfills. In addition, automation and smart manufacturing bring an enormous advantage to the industrial sector in terms of environmental and economic sustainability. Effectively, in traditional series production, possible defects are commonly detected at a later stage of production, and already with a large series of manufactured or semi-manufactured components, which brings an environmental burden and a much more significant cost for companies. The automation and democratization of detectors, artificial vision systems, machine learning, and big data contributed decisively to the fact that all the factors surrounding production control dropped significantly in price and could be easily integrated through more powerful computational systems, allowing real-time alerts and a much more assertive management performance, avoiding economic waste and undesirable environmental impacts.

Smart manufacturing also comprises numerous reconfigurable systems that make production much more flexible. These production systems are designed to avoid or minimize equipment setups. While conventional production is essentially based on conventional tools applied to unitary equipment, not always equipped with standard platforms that allow an easy exchange of these tools, smart manufacturing is designed in such a way that, if there is a need for tools, they are codified and strategically positioned, exchanged quickly, and avoiding significant loss of time. In this case, automation overlaps methodologies such as Single Minute Exchange of Die (SMED) and anticipates needs, creating flexible programmable systems that execute the necessary actions at the right time, avoiding human intervention, and corresponding in the most assertive way possible to the expectations created around the production schedule.

Alongside smart manufacturing, another very important concept has also emerged, called smart retrofitting. In fact, there are many millions of assets installed, which will last for years, and which do nothing to improve the environment, making it advisable to retrofit them, not just in the sense of making them more productive and competitive. Effectively, retrofitting decisions usually start from the need to make equipment or goods more productive, when one is talking about industrial equipment. However, environmental aspects also need to be considered. As the economic component overlaps the environmental component in most societies, it will certainly be necessary for many governments to establish programs to encourage the retrofitting of equipment or to implement punitive laws that oblige the industrial sector to rejuvenate its assets, with a view to complying with environmental goals. Smart retrofitting consists of consistently and targeted improving the hardware, software and connecting these assets to the Internet to improve their industrial and environmental performance, making the asset capable of being remotely commanded and monitored. This allows the equipment to be integrated into more elaborate production systems, in which production flows more efficiently and the logistical load normally associated with the transition of semi-products from process to process, or from equipment to equipment, is reduced. Jaspert et al. [5] studied the different definitions of smart retrofitting but, more importantly, summarized the main drivers and challenges associated with smart retrofitting.

In this study, the following were identified as smart retrofitting drivers: (a) ensure competitiveness; (b) increase equipment efficiency; (c) increase market responsiveness; (d) transition to smart manufacturing; (e) comply with regulations; (f) achieve sustainability. Among the 23 works consulted and understood as relevant to the study in question, the authors found that the papers were mostly focused on ensuring competitiveness (12 papers). Six of these papers focused on increasing equipment efficiency. Four also mentioned the need to increase market responsiveness. Two were studying smart retrofitting in order to transition to smart manufacturing. Two focused on compliance with environmental regulations, and only two mentioned sustainability as the main objective for conducting smart retrofitting processes. These results agree with what was mentioned earlier in this chapter, which can be commonly observed in countless industrial companies, mainly in developing countries: the economic issue almost always overlaps the social issue, and this also overlaps with the environmental issue. This situation is reflected in the non-compliance with the targets related to the reduction of emissions, an issue that has been discussed in countless summits between the most developed countries, but which has resulted in a huge failure. This situation, which has been going on for decades, calls for bold political measures, which politicians avoid for fear of losing popularity. If some more developed societies are more aware of climate change and know how to act in order to fulfill their duty in environmental terms, other countries with less favored levels of environmental awareness continue in a blind search for competitiveness, not looking at the means to achieve its ends: to provide better living conditions to its populations, even if they are mortgaging their future in environmental terms. Regarding challenges, the same authors identified the following: (a) limited resources, (b) high complexity, (c) diverse requirements, and (d) cultural issues. The overwhelming majority (15) mentioned the limitation of resources as their main challenge, which implies that governments act in this sense, increasing support for projects of this nature. Complexity is also mentioned as one of the main challenges (10), which indicates lack of knowledge and training. In this sense, governments, and business associations, may take a decisive action to change this situation. The last two challenges pointed out above seem not to be strongly referred to as challenges (four and two references, respectively), but they are nonetheless important. In fact, cultural issues are closely linked with the perception of difficulty adopting smart retrofitting, as discussed earlier. In fact, there are countries where the global mindset is focused on improving the environment, without jeopardizing their economic profitability. They know how to combine both needs, having a clear perception that environmental degradation can constitute an extremely serious problem for nature and for humanity, which is already being noticed through some serious climatic phenomena, such as the melting of glaciers and the average temperature rise on the planet. Other societies, because they have a very different economic reality, are much more focused on reducing poverty rates, regardless of the cost, without the environment being minimally considered. From these conclusions, some ideas should be kept in mind: education is something fundamental in the entire process of evolution, and awareness of the phenomena that surround us is vital to structure our thinking about the strategic decisions that we should take now, with a view to achieving a better future.

After this more general introduction, it is now time to study how smart manufacturing can contribute in an extremely positive way to a better environment.

5.2 Process improvement

On a planet where the population continues to grow vigorously, where global economic power has also been registering a relatively positive evolution, where energy is increasingly necessary in our daily lives, and where fossil fuels are still one of the main energy resources, all efforts to rationalize energy consumption are welcome. These efforts have drawn the attention of numerous researchers who, more occasionally or more generally, have been disseminating strategies to make rational energy consumption a reality.

Industrial processing, in most cases, is a strong consumer of energy (about 57% of the total energy consumed in China [6]). Moreover, it is also a strong emitter of greenhouse gasses, and in certain cases, it is responsible for the emission of polluting gasses liquid effluents that are not always properly treated, and waste solids that are difficult to recycle and that are deposited in landfills. In China, the industry is responsible for approximately 70% of the total carbon emissions [7,8]. The definition of the product, and the definition of processes, can play a major role in reducing the harmful effects on the environment described above. More than just reducing pollution, it is necessary to combat it vehemently and assertively through concerted policies and actions sustainable over time. It is common for some researchers to state that digitization makes a decisive contribution to reducing the environmental footprint attributed to industry. If the statement can be considered true, it is also necessary to consider the impact that digitization has on this environmental footprint, and perhaps it is not as significant as advertised. The adoption of smart manufacturing can be a valuable help in reducing energy consumption, and if combined with the definition of the product, it can significantly reduce the use of some resources, thus contributing to a smaller ecological footprint attributed to the industry.

5.2.1 Process industry: the case of chemical industry

The chemical industry is normally integrated into what is conventionally called the 'process industry', such as the petrochemical industry and the industry of detergents, paints, solvents, among many others. It is comparable to the steel industry, building materials, and biomedicine , among many others. Basically, all processes are linked and the production flow is practically continuous. Feng Qian [9] succinctly described , through a diagram (Fig. 5.1), how the processes are integrated between the necessary resources for production and the final use, divided by several main sectors.

The process industry constitutes a pillar of unparalleled importance for the economy and social development of any country and has been one of the growth supports of most developed countries. However, this type of industry also presents some important challenges [9].

- The constant variation in the properties of raw materials requires a continuous adaptation to continuous processes. It is quite complicated to characterize, adjust,

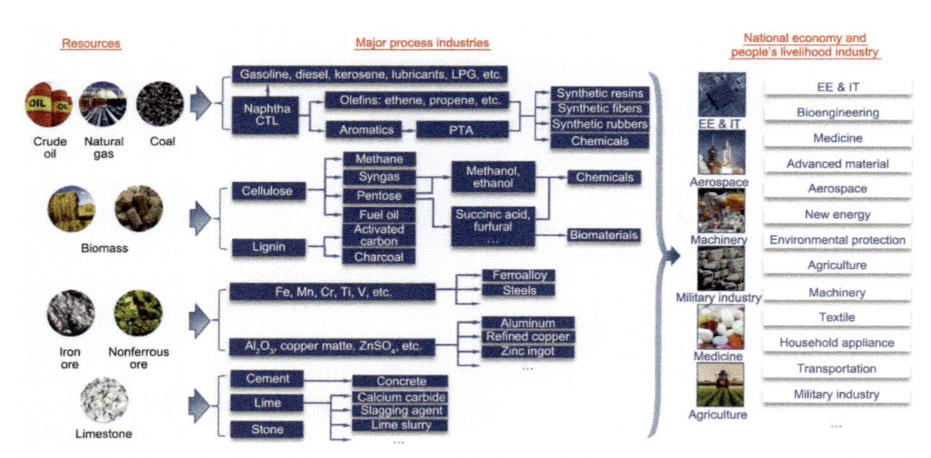

Figure 5.1. Schematic of flow in the main process industries, which comprise a large part of the chemical industry [9].

and perform stationary control in multiphase and multicomponent industrial processes.

- The complexity in increasing production rates by increasing the capacity of the equipment coupled to the process is very large because the reactions do not occur in the same way, changing the scale. Adding to this the lack of homogeneity in the characteristics of the raw material that feeds the production system, it becomes quite difficult to have an absolute mastery of the characteristics of the final product.
- Considering an open environment but under strict restrictions, economic control, process stability, and monitoring are huge challenges when aiming at multi-objective regulation of manufacturing processes in the process industry.

However, these problems can be partially or completely solved by introducing several tools associated with smart manufacturing, such as IoT, Big Data, Cloud Manufacturing, Artificial Intelligence, and Machine Learning.

Precisely for this purpose, Mao et al. [10] conducted research to list the main challenges of the process industry using artificial intelligence. First, they identified three categories of the main problems related to the process industry and green manufacturing: (a) integration of information; (b) real-time risk assessment and adequate support in decision making; (c) early warning. Based on these problems, a list of challenges was generated and explained in more detail below:

- **Acquisition of knowledge in very unique environments:** the acquisition of knowledge requires a sufficiently large variety of situations to allow knowing the boundary between what is expected and what is not expected. In the process industry, sampling is very relative, as the input to the system is constant, and if the process does not experience fluctuations, the output will be equally invariant. Thus, knowledge becomes something restricted. However, the chemical industry can generate an almost infinite number of reactions, and a small fluctuation in the input variables, or in the process, could generate different characteristics in the product obtained, or create safety problems that could be serious enough to put human lives at risk, in addition to having a serious influence on the environment. Furthermore, the chemical industry, despite appearing to be confined to a scientific area, is extremely multidisciplinary, requiring medium- or high-level knowledge in many other areas. The programming of a process, in addition to knowledge in the area, requires deep knowledge of automation, equipment, and human action. In fact, a perfect symbiosis between all this knowledge is required. A failure can generate serious incidents, and unfortunately, history reports some very serious cases of safety failures in the process industry related to the chemical industry, which resulted in significant loss of life and significant environmental damage. To minimize security breaches, the system must be programmed to monitor the most important variables and issue alerts if these variables exceed control values. Thus, the knowledge required to generate the procedure for the process and safety control requires levels of knowledge that are difficult to obtain immediately and that are not easy to train due to the specificity and scope of each process.

- **Reasoning supported by knowledge about process safety:** because risk mitigation and safety in the process industry should be a priority, it is necessary to have a database as wide as possible on anomalies, conducting extremely frequent screening of all conditions that could constitute breaches of security for the production system. This monitoring needs to be carried out preferably in real time and, above all, be reliable. Proper programming of algorithms that aim to cross data in real time and the comparison of the results of these analyzes with reliable databases could allow monitoring systems to detect cause-effect relationships that would not be easily discernible for human beings, even when endowed with sufficient analysis experience in this type of situation. Thus, monitoring systems supported by knowledge (databases) could be extremely useful for avoiding security breaches. Literature states that the reasoning methods supported by the most recent knowledge can achieve accuracy in the order of 80% [9]. However, given the risks and effects of a possible failure, this level of certainty/uncertainty is unacceptable. Therefore, there is still a clear opportunity and need for improvement here, with a view to minimizing the safety risk in the process industry and satisfying the requirements that it imposes.
- **Combining heterogeneous data coming from different sources:** in the process industry, data related to process safety arises based on different situations: static and dynamic. Data corresponding to the static situation comprise process information and risk analysis documents that are previously known and rarely updated, whereas dynamic data mainly describe the state of the process, which has its own dynamics and needs to be evaluated in real time. In practice, it is common knowledge that the data obtained incorporate ambiguities, which negatively interfere with the acquisition of knowledge. This situation can be overcome in two different ways: the first focuses on data pre-processing, while the second emphasizes the acquisition of knowledge based on a domain, combining in an appropriate way the knowledge through the acquired data.

- **Effective learning strategies for dynamic risk assessment and support in decision making:** preferably, the cause and effect relationships in a chemical process are restricted to a group of entities, complying with certain previously defined rules or axioms. Although the accumulated knowledge allows the analysis of the cause and effect and of specific deviations, there is usually some complexity in defining what the acceptable deviation is. Ideally, the process should allow real-time collection of data that the process monitoring system can process to identify possible deviations from the state established as normal. However, knowledge of the process must be supported by machine learning methodologies that allow quick identification of anomalous situations. Then, with a properly structured algorithm, the reception of certain deviations from the normal state can be evaluated at different stages throughout the process, allowing to define in an assertive way what the possible consequences will be, that is, what characteristics the final product may have, take, or make an analysis of the risks that are being taken or may occur. Finally, with the dynamic risk assessment performed for each series of events, a final decision can be made. In any case, to arrive at this decision-making, vast information is necessary. The reliability of collecting data from process monitoring is critical. There must be perfect knowledge of the failure modes of each piece of equipment and corresponding effects, as well as operational procedures to mitigate them. Unfortunately, in practice, high quality data are far from sufficient to guarantee adequate learning of the algorithm. Starting from a small set of data and extrapolating them to an algorithm that guarantees successful operation in the first stages of production will always be a challenge.

In fact, both AI and other components of smart manufacturing can contribute to a more sustainable, greener, and safer chemical industry, which has been a trend in the last decade. In the chemical industry, the more assertive and real-time monitoring of processes led to the realization that a continuous flow invariably led to savings in material resources and energy, assuring even a lower incidence of risk for the process, thus increasing safety [11]. Thus, this trend has remained constant recently. In addition, closer monitoring of a continuous process makes it easier to understand whether there is room for multidimensional reactions, which would, in turn, increase performance, process design, and efficiency. This improvement can generate a chain reaction that allows the development of state-of-the-art equipment capable of ensuring lower process risks, both in terms of safety itself and in terms of the fidelity of the characteristics of the product intended to be placed on the market. These developments could, in turn, trigger new predictive technologies for reactions, allowing greater control and automation of processes, ensuring that alerts would be issued at the right time to avoid degeneration of product characteristics and/or reactions that could present safety problems for the facility, its workers, and the environment. However, and according to the concept of green chemical industry introduced by Paul Anastas and John Warner in 1998 [12], in order to make the chemical industry follow a greener path, it is necessary to use nobler ingredients and fine chemicals, and an effort in that direction since then has been noticed.

However, it must remain clear that maintaining the performance of a reaction across multiple reactor scales and configurations requires relatively consistent dimensionless amounts for heat transfer, mass transfer, mixing, and reaction rates. Effectively, there are situations that are completely undesirable, such as moving from an operating regime in which the mixture is properly carried out to another in which there are transport limitations, generating undesirable temperature gradients, or high concentration gradients, generating unwanted by-products. The medicine-producing industry's main objective is to increase its efficiency, designing product continuous flows, avoiding waste and intermediate steps. It integrates the previously mentioned principles of smart manufacturing, in which the previously described integration of equipment is replaced by the integration of reactors, each with its specific function. The most common approach to increasing efficiency in a continuous production is to drastically reduce waste. The continuous production flow incorporates mixing, reaction, separation, and purification modules, a situation that is completely different from batch production, with each of the previous operations being carried out on a certain portion of feedstocks sequentially, but individualized by operation, and not continuously. Another way to increase efficiency is to conduct streams of by-products that have not reacted and direct them immediately to recycling, i.e., rework on the same by-products, avoiding losses. With the increasing introduction of automation in the continuous processes of the chemical industry, both for drugs and other chemical products normally produced continuously, reconfigurable flow reactors have emerged that allow the continuous processing of certain necessary chemical products by automatically selecting the most efficient path (reactions sequence), and greener, using sustainable feedstocks and environmentally friendly solvents. This step is followed by autonomous modification/reconfiguration of the multistage reaction platform, resulting in safer process operation and almost negligible waste production.

Becker et al. [13], in a recently published review study, listed the principles and regulatory guidelines that should govern the green chemical industry, with a particular focus on the pharmaceutical industry. Table 5.1 replicates the summary published by Becker et al. [13]. To avoid cross-contamination of pharmaceutical products and by-products, target values of acceptable daily intake and maximum allowable carry-over of a compound to the next product, defined for each active pharmaceutical ingredient grounded on non-health-based exposure limits, should be exceeded. Thus, for batch reactors and equipment, intermediate cleaning and before changing products will have to be performed until these limits are reached, which normally requires careful and time-consuming cleaning with organic solvents and water. This factor is especially critical in the development stages, where only a few experimental batches are manufactured at each stage of the process. Given that the environmental impact generated in the development phase of a pharmaceutical product is considered relatively small when compared to series production, increasing the use of solvents for cleaning the reactors during the development phase can change the assumption of low environmental impact at that stage of development. However, and from a perspective of more general use of solvents, there are recent investigations that refer to the use of surfactants in water to form nanoreactors in organic media, inducing conventional organic reactions to be performed in water. This strategy has

Table 5.1. Green chemistry principles and selected regulatory guidelines.

Green Chemistry Principles	Regulatory Guidelines
Prevent waste instead of treating it	ICH Q11 - Development and manufacture of drug substances
Design of atom-efficient synthetic methods	ICH Q3C - Residual solvents
Choose synthetic routes using nontoxic compounds where possible	ICH M7 - Assessment and control of DNA reactive impurities in pharmaceuticals to limit potential carcinogenic risk
Design new products that provide functionality while reducing toxicity	ICH Q8 - Pharmaceutical development
Minimize the use of auxiliary reagents and solvents	Cleaning procedures to avoid cross contamination
Design processes with minimal energy requirements	FDA guidance advancement on Emerging Technology Applications
Preferable use of renewable raw materials	ICH Q3A/B - Impurities in new drug substances/ products
Avoid unnecessary derivatization	ICH Q6A - Specification and acceptance criteria for new drug substances and new drug products: chemical substances
Replace stoichiometric reagents with catalytic cycles	ICH Q1A –F: Stability of drug substances and drug products
Design of new products with biodegradable capacities	FDA Guidance Quality Considerations for continuous manufacturing
Develop real-time and online process analysis and monitoring methods	ICH Q9 - Quality risk management
Choose feedstocks and design processes that minimize the chance of accidents	

ICH – International Council for Harmonization (Technical Requirements for Pharmaceuticals for Human Use)

been followed with great success, with a view to eliminate or greatly reducing the use of environmentally friendly solvents. The results are still not completely conclusive, but in preliminary trials of the drug TAK-954, a dosage of 0.5 mg/day with the new green solvent has been shown to have the same effect as another previously approved by the Food and Drug Administration (FDA) of the United States drug using conventional solvent with a dosage of 10 mg/day, i.e., a 20 fold greater effect [14]. Looking at sustainability from the perspective of drug development and manufacturing in the chemical industry, it is vital to monitor the amount of waste or reagents generated by each kg of a given product produced, known as the E factor and Process Mass Intensity (PMI). These indices are widely known and provide performance indicators capable of translating and evaluating the effectiveness and sustainability of the process. However, this metric by itself, may not reflect the entire sustainability of the process and the product. In fact, and as previously mentioned, it is necessary to quantify the effect of the product and the effluents generated and their toxicity. It is necessary to analyze the product as if doing an analysis of its life cycle, since a much smaller amount of one drug can produce the same effect in the patient as a large amount of another, and/or produce a much smaller amount of toxic products as waste. Therefore, the analysis is much more complex than analyzing just one indicator, but it must be carefully carried out with a view to moving in the true direction of a more sustainable and environmentally friendly chemical industry. To circumvent the referred complexity, new metrics were recently developed that define the degree of Relative Greenness of the Process (RPG), allowing the comparison of processes from the point of view of green chemistry [13]. However, this indicator may not serve the entire chemical industry, which is fragmented by sectors, and will still require further work to find indicators that adjust to all the needs of this industry.

Meanwhile, several studies have also been published that , in a more focused way, have allowed for the recording of advances in the green chemical industry, and which will now be described very briefly.

Miele et al. [15] successfully manufactured polydioxanone (PDO) microfibers using centrifugal spinning, a manufacturing process that uses less energy and is therefore considered green and more sustainable. The sustainability and low ecological footprint of this process is also reinforced by the fact that it does not use solvents harmful to the environment, having used Deep Eutectic Solvents (DES), such as plasticizers and betaine/citric acid or choline/citric acid (both in proportions of 1:1 by weight). The DES:PDO 20:80 ratio allows for an adequate process yield, and regardless of the DES, the microfibers have a smooth surface and diameters between 10 and 20 μm. The microfibers remained stable during the process, as confirmed by solid state analysis. Good resistance to compression (disk, random 3D arrangement) and tensile strength (rope, aligned 3D arrangement), as well as high elasticity and reduced deformation, were confirmed through mechanical tests. These properties are necessary for implant performance *in vivo*. Hydration practically does not affect fiber morphology and degradation is very slow, which supports homing and proliferation of fibroblasts. Furthermore, the microfibers are cytocompatible with human fibroblasts *in vitro* and have excellent safety and efficacy profiles in a live murine burn/excision model.

Viveiro et al. [16] proposed the development of a new generation of heterogeneous catalysts that are cheaper and less harmful to the environment. These catalysts, called TEMPO-MIP 10 (enzyme-inspired polymeric powder particles), have been shown to be very selective and fast in oxidizing cholic acid hydroxyl groups, which indicates a highly specific catalytic system for cholesterol derivatives in this molecular weight range.

In turn, Milovanovic et al. [17] developed an environmentally friendly process to obtain oil extracts from milk thistle seeds, a very valuable raw material in industrial terms, and studied the influence of process parameters. For this purpose, they used a supercritical green solvent, CO_2, designating the process as supercritical fluid extraction. With a view to analyzing the influence of the process parameters on the recovery of essential fatty acids and phenolic compounds, the authors carried out studies around the influence of the pressure needed in the high-pressure unit and temperature. The characterization of the high-value extracts obtained was carried out using a gas chromatograph, combined with a mass spectrometer, and a spectrophotometer. The authors stated that the process was influenced by (a) the origin of the seeds; (b) their year of harvest; (c) storage conditions; (d) nominal pressure of the equipment; (e) amount of mass initially introduced in the equipment; (f) parameters used in the process, such as temperature, pressure, and process time; (g) solvent/feed ratio. It was verified that the conditions that allowed a greater efficiency of the process were a pressure of 450 bars and temperature of 80ºC but, given that these conditions are harmful to a good functioning of the reactor over time, 300 bar and 60ºC were defined as possible alternatives.

Gluconic acid and its derivatives have high potential for application. Based on this, Pal et al. [18] performed a review on the evolution of the production process of gluconic acid and its derivatives, with a view to making it more efficient and environmentally friendly. The most traditional approaches to the production of organic acids are catalytic oxidation and fermentation. In any of these approaches, downstream purification involves several steps, such as centrifugation, carbon adsorption, evaporation, crystallization, and ion exchange. Given the importance of this product and its derivatives, producing it with high purity at an affordable price in an environmentally efficient way is a well established challenge, essentially due to the involvement of several downstream processing units and the associated energy consumption. This leads to relatively high costs. The possibility of integrating customized and highly selective membranes and modules with a fermenter in the downstream purification of gluconic acid provides some hope of reducing production costs without loss of purity. Furthermore, the process consumes significantly less energy, which is environmentally attractive. Thus, it appears that there are now alternatives to traditional processes that make the production system greener and more economical. New approaches include the use of membrane-based downstream purification of gluconic acid from fermentation broth.

Through some of the developments found in literature, it can be seen that the chemical industry, together with academia, are trying to make their processes more effective, safe, and environmentally friendly. However, the efforts found are more related to the pharmaceutical and food industries than to some other industries that are more dangerous and harmful to the environment, such as the production of fertilizers, insecticides, petroleum products, etc. The most dangerous and harmful chemical industry for the environment is concentrated in economically disadvantaged countries, where research is not a priority On the other hand, the richest and most developed countries feel that the danger is further away and that disadvantaged countries need to manufacture these products, with inherent risks, to minimize the social problems they suffer from. Therefore, it is not surprising that research focused on the most dangerous products and substances continues with a clear lack of focus on the part of the main research groups.

5.2.2 Additive Manufacturing as a booster of environmental sustainability

As mentioned earlier in a previous chapter, additive manufacturing is one of the most important pillars of the smart manufacturing group of processes. It consists of starting from a 3D model designed in a specific software, in which the model is transformed into slices by another specific software for this purpose (Slicer), which provides orders to a 3D printing equipment that fulfills the trajectories necessary to fill each layer defined by the Slicer, depending on the 3D model initially defined. This is a disruptive technology that has shaken all conventional manufacturing processes that have already been widely studied and debated, definitively eliminating some barriers placed on the project in the definition of some parts [19].

This manufacturing process also stands out for not requiring molds or tools and for being able to use numerous materials from different families, from polymers to composite materials and biomaterials. In this manufacturing process, more than a dozen different technologies are integrated, which address the same production philosophy, but using completely different means, maintaining the same principle: the components are built layer by layer. Having started with polymers, there is currently large research in the area of production of human organs using biomaterials [20]. Along the way, the development of prostheses was also a subject of investigation, mainly because it is possible to make a personalized survey of the geometry of a patient's organ through 3D scanning and faithfully reproduce that organ, thus drastically reducing the adaptation phase of a given model to the required morphology for the patient. This manufacturing process fits perfectly into what is called smart manufacturing, as despite still requiring some human intervention, the components can be designed, passed through the slicing software, and introduced into the production equipment. Only the post-processing steps may require more human intervention.

Additive manufacturing is clearly a green process because, except support constructions, or possible wrong impressions that lead to waste, the process concentrates all the energy and raw material in the realization of the component intended to be built. However, it still has other advantages over conventional processes. Given its construction philosophy, the design of new components can be an aggregate of components previously produced by conventional technologies. This also has environmental and lead-time implications. Effectively, the aggregation of different parts into a single one, even if they require relative movement between them, allows avoiding internal logistics operations, which results in less energy consumption and unequivocal time savings.

It can be said that the additive manufacturing process is an energy-consuming process. Some technologies used equipment whose performance is not the most efficient, namely, laser beams. However, the energy is concentrated at the same point, so the waste is not very high. It should also be mentioned that, considering some powder bed processes, mainly in the 3D printing of metallic components, there is some waste of energy through heat dissipation with each deposited layer. All of this being correct, it is also true that conventional processes waste a lot of energy, so it can be said without any doubt that this is a highly automated process and that it is environmentally friendly, saving raw materials and, in most cases, also saving energy, even considering post-processing operations. Given the sophistication of the process, communication, and IT resources used and the automation it provides, it can be considered perfectly integrated into the definition of smart manufacturing.

It is not the purpose of this work to describe the different technologies that can be used in 3D printing, but here a short review of some of the main processes is presented:

- **Stereolithography (SLA)** is a 3D printing process used primarily with polymers, which uses a laser beam to cure and solidify a photopolymer resin in layers. SLA is suitable for producing small parts with fine details and good accuracy.
- **The digital light processing (DLP)** process is like SLA and is based on the light curing process, being applied essentially with polymers. The SLA process uses a laser, whereas the DLP process uses light from a high-resolution digital projector to cure the photosensitive resin used in the manufacture of parts and prototypes.
- **The digital light synthesis (DLS)** process allows 3D printing of parts and models in a similar way to DLP-based SLA methods, where the build platform is submerged in a resin bath, and the cross-section of the object to be printed is projected by DLP through the existing transparent bottom of the resin tank. As the resin cures, the printing plate rises, bringing the printed part with it. The process is essentially used with polymer raw materials.
- **Selective laser sintering (SLS)** is a "powder bed" melting technology that uses a laser beam to locally sinter a thin layer of polymeric material that is presented in powder form and, thus, build parts layer by layer. After the fusion of a layer by the laser beam, a new complete layer of powder is distributed throughout the working area, and the laser starts melting a new layer of powder, and so on, until the end of the printing process.
- **Fused Deposition Modeling (FDM)** is also known as Fused Filament Fabrication (FFF) in the case of small equipment and is commonly known as desktop. The FDM process uses a coil of thermoplastic material that feeds an extruder head, which heats the material that is deposited through a nozzle layer upon layer, allowing the construction of the part. On most FDM printers, the nozzle moves in the X and Y directions while the build pad moves in the Z direction as the part is formed. This process works only with polymers.
- **Material jetting (MJ)** technology uses print heads to deposit molten material through jets that, when deposited on the layer, cool and solidify. The material jet process works like a standard inket printer, in which layers are built up by depositing material on top of each other to create a solid object. If the build space is optimized, the Material Jetting process can produce parts faster than other 3D printing technologies. This technology works essentially with polymer raw materials.
- **Binder jetting (BJ)** technology involves spreading a thin layer of powdered material over the built platform, followed by spraying a liquid binding agent applied via inkjet printheads, binding the particles together in the area delimited by the section of the component to be obtained. It works essentially with polymers.

- **The laminated object manufacturing (LOM)** process was initially developed by Helisys, Inc., but has now been discontinued because it is rarely used nowadays. This process allows the construction of solid models layer by layer, resorting to the deposition of sheets of material (usually paper sheets) joined together through heat and pressure, and where each sheet is cut with the aid of a laser beam or a blade. It was used mainly to shape prototypes of relatively large dimensions, such as car bumpers and dashboards. Nowadays, 3D printing technologies are faster and do not require the post-processing time that this technology requires.
- **Electron beam melting (EBM)** is a technology very similar to selective laser sintering , essentially intended for metals and metallic alloys, in which the raw material is used in the form of powder, being distributed layer by layer. After the distribution of this powder, the beam of electrons travels through the area corresponding to the solid part that needs to be built, merging the material of this layer with the material already solidified in the previous layer. In this process, like the others mentioned above, the components are built layer by layer.
- **Selective laser fusion (SLM)** is a technology with greater industrial and academic classification than Electron Beam Fusion (EBM), which uses one or more laser beams to promote the fusion of specific regions of the powder bed. These devices usually use fiber lasers, Nd:YAG or Ytterbium, with a wavelength of 1.06 μm, with different powers up to 1000 W, and laser spot dimensions between 50 and 100 μm. They usually have galvanometric mirrors that direct the beam over the available area, which affects the construction rates of this technology. The deposition chamber of the selective laser melting equipment works in an inert atmosphere, usually argon or nitrogen, to reduce the oxygen content and the risk of combustion of small metallic powders.
- **Direct energy deposition (DED)** uses an electric arc, electron beam, or laser as a source of energy, which is concentrated to promote the fusion of the material projected onto a substrate through a nozzle. Technologies suitable for this process do not melt a pre-deposited metal on a substrate, as in powder bed melting technology, but melt the metal as it is released from the deposition nozzle. This process promotes the creation of a melt pool, depositing material continuously, layer by layer. Using this process, it is possible to deposit metals in the form of powder or wire to create three-dimensional components, which have been used to manufacture steel, titanium alloys, cobalt alloys, nickel alloys, aluminum alloys, and even alloys with the memory of shape. Some DED technologies deposit the metal in a closed chamber with a controlled atmosphere or under vacuum (such as LENS technology (Laser Engineered Net Shaping)), while others use a flow of inert gas, usually argon, to create an atmosphere protection with reduced oxygen content to prevent oxidation of the part material (such as DMD technology (Laser-aided Direct-Metal/material Deposition)). In the case of manufacture using powder, there could be a carrier gas for the powder and a separate outlet for the protective gas, with different possible configurations for the nozzle and powder outlet.

- **Direct laser deposition (DLD)** technology uses high-power lasers as a source of thermal energy to create a molten pool on the surface of the substrate. Nd:YAG lasers and fiber lasers are usually used. This technology uses metal alloy powders, which are directed through deposition nozzles using a gas as a carrier, to melt under the influence of the laser. This technology promotes rapid melting and solidification of the metallic powder to manufacture the part layer by layer.

These are some of the most used processes, mainly for processing polymers, metals, and metal alloys. However, some of these processes can also be used with ceramic materials, composite materials, and even biomedical materials, which require extra care with the energy that is used, given the complexity and vulnerability of the materials. Although, apparently, some of these technologies involve a large volume of material, corresponding to the delimitation of a parallelepiped that covers the entire volume of the component, almost all excess material is reused, allowing one to state that this process is environmentally friendly and sustainable. In the case of processes that use a bed of powder, for example (they are not the only case), although successive layers of powder are spread, only the powder that corresponds to the geometry of the component is melted layer by layer, while the other powder in excess remains free and can be vacuumed for later reuse. Therefore, it can be said that practically only the material that is effectively needed to make the component is consumed.

In addition to the preceding sustainability attributes, additive manufacturing provides exceptional opportunities to review the entire design of the different components and their interaction, as it handles complex formats quite well. While traditional processes were limited to the possible geometries of the tools (casting, forming, machining, welding, injection), additive manufacturing builds the 3D model layer by layer, almost free of restrictions (deformations should be controlled through project optimization and use of support structures). Each layer, depending on the technology used, may have layers with a thickness of one or a few tens of microns, which allows for an excellent resolution, a factor that translates into moderate roughness, but not comparable to some machining processes, for the time being. Thus, given the design freedom granted by these technologies, it became possible to use much more complex shapes [21]. In addition, and using Finite Element Analysis (FEA), specialized techniques for optimizing components have emerged, considering the efforts to which they will be subject in work. This technology was named Topological Optimization, and its principle is based on using material only where it is completely necessary to comply with the efforts to which it will be subjected [19,22,23]. With this technology, the design easily achieves complex organic shapes that comply with the mechanical resistance requirements that are initially important by indicating to the software the places where the loads will affect and their intensity will be As in this case, the geometry is not dependent on a pre-produced shape (semi-products, such as angles, bars, tubes, etc.), the software freely designs the part, purging it of all material that would be useless in terms of resistance mechanics. With this, it can be said that in most cases the components lose weight, which is favorable for the environment in two ways: (a) when losing weight, especially in situations of mobility, the components will consume less energy for the whole where they are inserted, which corresponds to lower emissions; (b) the weight corresponds to material, so by having less weight, we are consuming less resources, which is also good for the environment.

The materials used in this process are normally quite expensive, and the production time is still relatively slow. However, some technologies have been studied, tested, and validated completely by the aeronautical industry, which even acquired some equipment factories and the respective technology, starting to produce parts that were very complex to produce, such as the turbine blades of jet aircraft. This process replaces lost wax casting, which requires very special care from the design to the final control. In environmental terms, it should also be noted that the process easily allows previously used raw materials that are intended to be recycled, which is an enormous asset [24].

Next, a summary of some research that has been carried out to demonstrate that additive manufacturing is a highly automated and quite a sustainable process.

Agnusdei and Del Petre [25] conducted a review and bibliometric study on the theme "Additive manufacturing for sustainability" and found that there is an almost exponential growth in the number of publications, mainly between 2017 and 2021, targeting this subject. Summarizing the conclusions, the authors also focused on optimizing the geometry/shape of the designed components, pointing to 'generative design' as a very useful way to reduce the use of material in the components, optimizing their shape according to the requests that will suffer in service, as it was previously reported with topological optimization. Aspects related to logistics are also focused on, as components can be manufactured on-demand, avoiding centralized manufacturing in some part of the planet, followed by time-consuming, expensive, and environmentally unfriendly transport to get this component to its destination. Some equipment manufacturers supply component drawings to customers so that they can reproduce them locally when a fault occurs, thus saving time and emissions with the transport of components between the products and the customer. In terms of management, this new production philosophy also contributes to the reduction or elimination of stocks, which reduces costs and makes processes more economically sustainable.

Calignano and Mercurio [26] also focused their attention on the logistical advantages of the process Effectively, this manufacturing process makes it possible to eliminate intermediate nodes in the supply chain, reducing transport time and bringing production closer to the final consumer. At the limit, and as previously mentioned, customers can become producers of the components they need or may need, eliminating the physical flow throughout the supply chain, with the respective advantages in environmental terms. Rinaldi et al. [27] repeated the same idea in environmental terms, which also contributes to economic sustainability. They also added that the flexibility in terms of production brought about by this process is extremely significant when compared with conventional processes. However, they pointed out that the main problem is that the production rate is relatively limited, negatively influencing the costs and competitiveness of the components manufactured based on this process.

Javaid et al. [28] also discussed the role of additive manufacturing in improving environmental sustainability, listing a significant set of applications where this process can be applied, bringing clear advantages to the environment. In addition, this study also focuses on some very interesting aspects related to additive manufacturing, noting that the production of lighter components will lead to energy savings during the useful life of the component, and it may extend the life of the components by using more adjusted alloys to the operating requirements to which the component will be subject, allows the faster replacement of any component in the case of failure, minimizing downtime and, above all, touches on a point that is also quite important: additive manufacturing allows the production of system modules, which can undergo upgrades that will benefit their operation, increasing their useful life. The aspects of reducing the logistics chain and the possibility of using recycled materials are also addressed.

Given that the Wire-Arc Additive Manufacturing (WAAM) process has gained significant importance in recent years, Mu et al. [29] decided to develop a framework to streamline the use of WAAM through the application of the Digital Twin. In addition to establishing a model that rivaled the frameworks already created for other additive manufacturing technologies, the work intended for this framework to be closer to green manufacturing, trying to significantly reduce post-processing operations and the need to conduct destructive testing.

In turn, Dritsas and Fernandez [30] developed a manufacturing process using exclusively natural materials, namely cellulose and chitin, deposited using an extrusion-based method. Natural biomaterials act differently than conventional plastics; therefore, forcing them to act like conventional plastics, for example, being thermoformed and resistant to oxygen and water, would not be the best strategy to extract all the potential of their ecological characteristics. However, they have unique properties that are worth exploring to learn how to best use them. These authors developed the FLAM 3D technology, which is based on the principles of extrusion, more accessible and widely available, will allow designers, engineers, and manufacturers to experience this process first-hand and gain knowledge and perception of the advantages of this process when combined with this type of material. The authors intended that this study will provide a better approach to the process, thus motivating the development of innovative products that are naturally sustainable.

Jayawardane et al. [31] conducted a review of frameworks developed with a view to relate additive manufacturing to sustainability. In addition to the factors already mentioned above, the authors draw some noteworthy conclusions, such as: (a) when comparing additive and subtractive manufacturing, it is necessary to consider the technical feasibility of being able to do it both ways, as the estimated useful life can significantly affect the life cycle results; (b) with a view to coordinating the study of environmental and economic impacts, the best approach to consider for a correct integration is the Eco-Efficiency Assessment (EEA); (c) to compare the social impact of both additive and subtractive manufacturing, Social Life Cycle Assessment (SLCA) and Human Toxicity Potential (HTP) assessment methods should ideally be used.

References

[1] Manco, P., Caterino, M., Rinaldi, M. and Fera, M. (2023). Additive manufacturing in green supply chains: A parametric model for life cycle assessment and cost. Sustainable Production and Consumption, 36: 463–478. doi: 10.1016/j.spc.2023.01.015.

[2] Butturi, M. A., Marinelli, S. and Lolli, F. (2022). A framework to assess the sustainability of additive manufacturing for spare parts. IFAC PapersOnLine 55-10, 1509–1514. doi: 10.1016/j.ifacol.2022.09.604.

[3] Lyu, W., Wang, T., Hou, R. and Liu, J. (2023). Going green and profitable: The impact of smart manufacturing on Chinese enterprises. Computers & Industrial Engineering, 181: 109324. doi: 10.1016/j.cie.2023.109324.

[4] Abudureheman, M., Jiang, Q., Gong, J. and Yiming, A. (2023). Technology-driven smart manufacturing and its spatial impacts on carbon emissions: Evidence from China. Computers & Industrial Engineering, 181: 109283. doi: 10.1016/j.cie.2023.109283.

[5] Jaspert, D., Ebel, M., Eckhardt, A. and Poeppelbuss, J. (2021). Smart retrofitting in manufacturing: A systematic review. Journal of Cleaner Production, 312: 127555. doi: 10.1016/j.jclepro.2021.127555.

[6] Liu, J., Yang, Q., Ou, S. and Liu, J. (2022). Factor decomposition and the decoupling effect of carbon emissions in China's manufacturing high-emission subsectors. Energy, 248: 123568. doi: 10.1016/j.energy.2022.123568.

[7] Liang, S., Yang, J. and Ding, T. (2022). Performance evaluation of AI driven low carbon manufacturing industry in China: An interactive network DEA approach. Computers & Industrial Engineering, 170: 108248. doi: 10.1016/j.cie.2022.108248.

[8] Kang, H. S., Lee, J. Y., Choi, S., Kim, H., Park, J. H., Son, J. Y., Kim, B. H. and Noh, S. D. (2016). Smart manufacturing: Past research, present findings, and future directions. International Journal of Precision Engineering and Manufacturing-Green Technology, 3: 111–128. doi: 10.1007/s40684-016-0015-5.

[9] Qian, F. (2023). The future of smart process manufacturing. Engineering, 22: 20–22. doi: 10.1016/j.eng.2022.04.029.

[10] Mao, S., Wang, B., Tang, Y. and Qian, F. (2019). Opportunities and challenges of artificial intelligence for green manufacturing in the process industry. Engineering, 5: 995–1002. doi: 10.1016/j.eng.2019.08.013.

[11] Zhou, Z., Qi, G. and Glimm, B. (2016). Exploring parallel tractability of ontology materialization. In: Proceedings of the 22nd European Conference on Artificial Intelligence; 2016 Aug 29–Sep 2; Amsterdam, the Netherlands. Amsterdam: IOS Press, p. 73–81. doi: 10.3233/978-1-61499-672-9-73. Available on: https://www.uni-ulm.de/fileadmin/website_uni_ulm/iui.inst.090/ Publikationen/2016/ZhQG16a.pdf. (Accessed on January 16th, 2023).

[12] Bennett, J. A., Campbell, Z. S. and Abolhasani, M. (2019). Role of continuous flow processes in green manufacturing of pharmaceuticals and specialty chemicals. Current opinion in Chemical Engineering, 26: 9–19. doi: 10.1016/j.coche.2019.07.007.

[13] Becker, J., Manske, C. and Randl, S. (2022). Green chemistry and sustainability metrics in the pharmaceutical manufacturing sector. Current opinion in Green and Sustainable Chemistry, 33: 100562. doi: 10.1016/j.cogsc.2021.100562.

[14] Chapman, M. J., Jones, K. L., Almansa, C., Barnes, C. N., Nguyen, D. and Deane, A. M. (2021). Blinded, double-dummy, parallel-group, phase 2a randomized clinical trial to evaluate the efficacy and safety of a highly selective 5-hydroxytryptamine type 4 receptoragonist in critically ill patients with enteral feeding intolerance. Journal of Parenteral and Enteral Nutrition, 45: 115–124. doi: 10.1002/jpen.1732.

[15] Miele, D., Nomicisio, C., Musitelli, G., Boselli, C., Cornaglia, A. I., Sànchez-Espejo, R., Vigani, B., Viseras, C., Rossi, S. and Sandri, G. (2023). Design and development of polydioxanone scaffolds for skin tissue engineering manufactured via green process. International Journal of Pharmaceutics, 634: 122669. doi: 10.1016/j.ijpharm.2023.122669.

[16] Viveiros, R., Maia, L. B., Corvo, M. C., Bonifácio, V. D. B., Heggie, W. and Casimiro, T. (2022). Enzyme-inspired dry-powder polymeric catalyst for green and fast pharmaceutical manufacturing processes. Catalysis Communications, 172: 106537. doi: 10.1016/j.catcom.2022.106537.

[17] Milovanovic, S., Lukic, I., Kamínski, P., Dębczak, A., Klimkowska, K., Tyskiewicz, K. and Konkol, M. (2022). Green manufacturing of high-value extracts from milk thistle seeds: Parameters that affect the supercritical CO_2 extraction process. Journal of CO2 Utilization, 63: 102134. doi: 10.1016/j.jcou.2022.102134.

[18] Pal, P., Kumar, R. and Banerjee, S. (2016). Manufacture of gluconic acid: A review towards process intensification for green production. Chemical Engineering and Processing, 104: 160–171. doi: 10.1016/j.cep.2016.03.009.

[19] Silva, F. J. G., Bartolomeu, F., Pereira, A. B., Duarte, T., Lino, J., Miranda, G., Neto, R., Relvas, C., Santana, L. and Terroso, M. (2023). Impressão 3D – Tecnologias, Materiais e Aplicações (in Portuguese). Engebook/Quântica Editora, Porto, Portugal. ISBN: 978-989-910-159-3.

[20] Arif, Z. U., Khalid, M. Y., Noroozi, R., Hossain, M., Shi, H. T. H., Tariq, A., Ramakrishna, S. and Umer, R. (2023). Additive manufacturing of sustainable biomaterials for biomedical applications, Asian Journal of Pharmaceutical Sciences, 18: 100812. doi: 10.1016/j.ajps.2023.100812.

[21] Hegab, H., Khanna, N., Monib, N. and Salem, A. (2023). Design for sustainable additive manufacturing: A review. Sustainable Materials and Technologies, 35: e00576. doi: 10.1016/j.susmat.2023.e00576.

[22] Silva, F. J. G., Campilho, R. D. S. G., Gouveia, R. M., Pinto, G. and Baptista, A. (2018). A Novel Approach to Optimize the Design of Parts for Additive Manufacturing. Procedia Manufacturing, 17: 53–61. doi: 10.1016/j.promfg.2018.10.012.

[23] Kumar, A. and Chhabra, D. (2022). Adopting additive manufacturing as a cleaner fabrication framework for topologically optimized orthotic devices: Implications over sustainable rehabilitation, Cleaner Engineering and Technology, 10: 100559. doi: 10.1016/j.clet.2022.100559.

[24] Wu, H., Mehrabi, H., Karagiannidis, P. and Naveed, N. (2022). Additive manufacturing of recycled plastics: Strategies towards a more sustainable future. Journal of Cleaner Production, 335: 130236. doi: 10.1016/j.jclepro.2021.130236.

[25] Agnusdei, L. and Del Prete, A. (2022). Additive manufacturing for sustainability: A systematic literature review. Sustainable Futures, 4: 100098. doi: 10.1016/j.sftr.2022.100098.

[26] Calignano, F. and Mercurio, V. (2023). An overview of the impact of additive manufacturing on supply chain, reshoring, and sustainability. Cleaner Logistics and Supply Chain, 7: 100103. doi: 10.1016/j.clscn.2023.100103.

[27] Rinaldi, M., Caterino, M., Fera, M., Manco, P. and Macchiaroli, R. (2021). Technology selection in green supply chains - the effects of additive and traditional manufacturing. Journal of Cleaner Production, 282: 124554. doi: 10.1016/j.jclepro.2020.124554.

[28] Javaid, M., Haleem, A., Singh, R. P., Suman, R. and Rab, S. (2021). Role of additive manufacturing applications towards environmental sustainability. Advanced Industrial and Engineering Polymer Research, 4: 312–322. doi: 10.1016/j.aiepr.2021.07.005.

[29] Mu, H., He, F., Yuan, L., Commins, P., Wang, H. and Pan, Z. (2023). Toward a smart wire arc additive manufacturing system: A review on current developments and a framework of digital twin. Journal of Manufacturing Systems, 67: 174–189. doi: 10.1016/j.jmsy.2023.01.012.

[30] Dritsas, S. and Fernandez, J. G. (2022). Towards sustainable additive manufacturing using Fungus-like adhesive materials. Materials Today: Proceedings, 70: 418–424. doi: 10.1016/j.matpr.2022.09.279.

[31] Jayawardane, H., Davies, I. J., Gamage, J. R., John, M. and Biswas, W. K. (2023). Sustainability perspectives—a review of additive and subtractive manufacturing. Sustainable Manufacturing and Service Economics, 2: 100015. doi: 10.1016/j.smse.2023.100015.

Chapter 6

Additive Manufacturing as a Greener Way to Production Systems

6.1 Principle, definition, and main features

Until a few decades ago, mechanical manufacturing processes were limited to what are now called conventional processes: casting, forming, welding, machining, injection, and plastic extrusion. The beginning of rapid prototyping in the 80's gave the first steps in a technology that later came to assume two different designations, but which converged in the same result: additive manufacturing or 3D printing. This technology represents one of the greatest evolutions in the application of automation to manufacturing processes. Rapid prototyping includes techniques that are not purely additive, such as Laminated Object Manufacturing (LOM), where after successive layers of paper and glue are added, all excess material has to be extracted, as the deposited layers are integral, subsequently generating waste that could be greater or lesser, depending on the fit of the shape to the standard volumetry generated [1]. Thus, the environmental sustainability process is not one of its main features.

However, the principle was defined, and it just needed to be perfected. The principle was based on depositing successive layers of material, the thinner the better, only in places that corresponded to the physical space to be occupied by the component to be manufactured. As can easily be perceived, the principle of the process differs significantly from all others existing until then, many of which were based on semi-products with established preforms, such as plates, bars, tubes, etc., regarding the specific shape of the needed product. Invariably, these so-called conventional processes generate waste (shavings, burrs, etc.), which, in addition to constituting an expenditure of excess material, imply an increased effort in recycling through significant energy consumption. The principle of additive manufacturing is much more environmentally sustainable because it only deposits material where it is needed to form the component. Obviously, the components can have a complex shape, which implies the use of support structures, representing some waste of resources, but it is significantly lesser than a conventional process.

Additive manufacturing is based on a series of steps that need to be fulfilled to reach the desired objective. The 3D model is generated using 3D CAD software and then exported, usually in STL format, whose acronym can be defined in several ways: (a) stands for stereolithography; (b) standard triangle language; or (c) standard tessellation language. The first acronym was defined by the company 3D Systems, which created and developed the stereolithography process. The file is exported to another software, usually called a "slicer", as it is responsible for converting the solid imported from the 3D CAD into slices as thin as the software and manufacturing equipment allow. Each slice generated by the "slicer" will allow controlling the equipment so that, through different technologies, depending on the variant in which people are working, the material previously deposited or that filled a certain space, can be solidified, and give rise to a new layer of the component to be manufactured. In this way, the component or set of components is built layer by layer, depending on the instructions issued by the "slicer", which obeys the solid previously generated by the 3D CAD software.

As mentioned above, this component construction technique is completely disruptive compared with conventional technologies. However, this is not an advantage. Indeed, there are factors that make this component construction technique extremely attractive, which can be listed as follows:

- **Generates extremely complex shapes:** Traditional processes have obvious and well-known limitations in terms of generating more complex shapes. Often, the required shapes and the material needed to meet the requirements demanded by the function of the component or the customer's needs dictate the conventional process that could be used, almost always existing constraints on the material and/or the manufacturing process adopted. It is usually said that there are no limits to the possible ways of creating and producing complex shapes in additive manufacturing. All organic shapes that were only possible to produce using certain processes such as casting or injection have virtually no restrictions in additive manufacturing. It may be necessary to use support structures to avoid deformations during construction, which represent an additional expense of material, but it is possible to build practically any shape, giving designers complete freedom of creation. In addition to the possibility of creating extremely complex shapes, there are still two enormous advantages in terms of shape: (a) the components can assume filling factors of less than 100%, and (b) assemblies can be manufactured, which, through conventional processes, would imply an assembly between components, and the fabrication of different components using the same or even different processes. As for fulfillment, this represents a remarkable evolution in the field of component design, its function, and its environmental impact. Certain components should have a certain shape and meet certain dimensions; however, the resistance they should have is small regarding the global dimensions of the component. However, through conventional processes, it would be impossible, or at least very difficult, to manufacture hollow parts or remove mass from the components. When additive manufacturing is used, the production equipment makes it possible to select which infill to use, creating a reticulated mesh that, while not having the same mechanical strength as the fully filled component, has the necessary resistance for the intended application. Obviously, this ability provided by additive manufacturing allows weight to be removed from the components, which translates into resource savings and less effort to recycle a posteriori. In traditional production, it is usual to manufacture each component, given the shape and processing difficulties in manufacturing, and then perform assembly operations. These operations demand manpower, time, and the consumption of components and can lead to assembly errors or quality problems. The precision patented by additive manufacturing makes it possible to manufacture components that are already integrated into each other, with a tenuous border of freedom between them if relative movement between these same components is necessary. If this is not necessary, the process makes it possible to add components that were previously manufactured individually, due to the limitations of conventional processes, and integrate them into a single component, again saving resources, time, and risks. Given that shape complexity is not a problem, there are no constraints in aggregating several components into a single component, facilitating the management of spare parts, stock and reducing assembly effort. Furthermore, using topological optimization software, by defining the support points and the loads that will be applied to a given component, it is possible to optimize its shape and filling through Finite Element Analysis (FEA) [2]. This software uses several iterations and can consider different loading conditions, providing solutions that would be difficult for the designer to achieve in a first approach, leading to solutions usually with very organic shapes and allowing a substantial reduction in the weight of the component. This solution is particularly advantageous for mobility applications, which is why it is largely used in the aeronautical and automotive industries, leading to lower consumption of fossil fuels. This solution is also extremely advantageous for the environment, as it enables a significant reduction in the resources used in the manufacture of components, without losing performance,

- **Work with a large range of materials:** Rapid prototyping began by using superimposed layers of paper, with glue deposited between these successive sheets, or adhesive sheets in which the glue was activated by a heated roller, and the shape was given by a relatively low-power laser that cut the sheet with the intended geometry for each layer. With the development of other technologies for rapid prototyping, other materials began to be considered and used. The Fused Deposition Modeling (FDM) technique, also known as Fused Filament Fabrication (FFF), was one of the first to be developed; however, because it was patented by the company that developed it, Stratasys, it only experienced real development when the patent expired, having democratized its use. As it is a very simple manufacturing system, in which the main role is in the equipment control software, there are currently systems for domestic use, given the possibility of acquiring equipment kits to assemble this type of equipment with costs being extremely low. This technology essentially works with thermoplastics such as polylactic acid (PLA), Acrylonitrile Butadiene Styrene (ABS), polyethylene terephthalate glycol (PETG), polyethylene terephthalate (PET), High-Impact Polystyrene (HIPS), thermoplastic polyurethane (TPU), and Aliphatic Polyamides (PA). However, technological development has led to the development of many other additive manufacturing technologies, and the spectrum of polymeric materials has grown exponentially, including photosensitive polymers, improving execution speed and component dimensional accuracy produced. The use of biomaterials is also already a reality, making it possible to reproduce human and other animal organs. However, other more ambitious steps are yet to be taken with the use of this technology in metallic alloys. The evolution of additive manufacturing techniques enabling higher powers and a uniform and very fine distribution of metallic powder, such as Selective Laser Melting (SLM), Laser-Powder Bed Fusion (L-PBF), and Direct Energy Deposition (DED), has allowed the evolution of complex metallic alloys. It became necessary to produce extremely fine and uniform metallic powders, but this also allowed powders of different elements to be mixed, forming alloys that until now were difficult to obtain through simple melting. Powder metallurgy paves the way for more complex alloys. Industries such as aerospace and aeronautics, with very specific needs, now have excellent solutions for applications that have used less sophisticated materials and more conventional manufacturing processes. This opened the way for manufacturing through additive manufacturing of components that, until now, used conventional technologies that were very likely to lead to defects, such as casting. General Electric (GE) found a good solution for the manufacture of components for aircraft propellers, creating a company from scratch with a view to using this technology in the manufacture of the required components. However, if it was possible to join metallic powders, it would certainly also be possible to promote the sintering of composite materials. For this purpose, Selective Laser Sintering (SLS) technology was used, allowing for another significant step in the development of additive manufacturing. The continuous development of technology allowed the distribution of powders to also be controlled, which allowed components to have different characteristics in different areas, thus giving rise to graded materials. Currently, it can almost be stated that there are no limitations to the use of the most varied types of materials, it being only necessary to select the most appropriate technology for each type of material. This is undoubtedly yet another extraordinary advantage of this technology, as unlike so-called traditional processes, it does not create strong constraints on the use of certain types of materials. The use of certain types of materials in specific applications can make a decisive contribution to saving materials, as more specific materials can be used for each need, more effectively covering the needs presented by each product. The use of less material, even if it is more expensive, reduces the use of resources, increasing environmental sustainability, even if economic sustainability is jeopardized.

- **Enables the generation of parts with different degrees of density/filling:** As mentioned in the earlier point, infill is an attribute that can be selected when manufacturing components. A 100% infill corresponds to a filled component, perfectly solid and whose component weight is directly proportional to the volume and density of the material used. Infills of less than 100% could be selected in situations where the part is not heavily stressed, where there are not particularly critical wear requirements, and where weight may be a critical issue. There are numerous situations where it is not necessary to use a 100% infill, which will result in less resource consumption and, if it is a part that will be used in mobility equipment, it represents much more than that for the environment, as it will allow significant energy savings throughout the life cycle of the components. Thus, this is a critical aspect to consider for the environment. Whenever possible, and if it does not compromise the function of the component or assembly, the infill must be heavily weighted and the lowest possible value used. This translates into savings in several areas, such as: (a) raw material savings; (b) energy savings in component manufacturing; (c) energy savings in moving the part in service (if applicable); (d) time savings in manufacturing; (e) less effort in recycling, or less ecological footprint if it goes to landfill.

- **Despite not having a perfect surface, the generated shape is very close to the final shape:** The deposition of successive layers induces some roughness on the external surface of the generated components. The thinner these layers are, the more definition the external surface of the components will have, i.e., the lower their surface roughness. However, the results are not linear. Depending on the material, the technique used, and the geometry of the component, the roughness could be affected. In the case of FDM technology, if the parameters are not properly defined, if the material melts too much, and if geometric contours are suspended or too organic, the roughness may be impaired. The effect of the layers during the manufacturing process is unavoidable, being felt more in some processes that are more sensitive to this parameter and less in others. As this problem affects the appearance and even the functionality of the components in some cases, solutions were quickly sought to overcome this problem. So-called post-processing techniques were then developed, which help to overcome the problem of roughness and can also be adopted for other purposes, such as improving mechanical strength or hardness characteristics, reducing residual stresses Post-processing techniques are primarily defined by the material of which the components are made. Processing techniques for polymers are substantially different from those used for metallic components or composite materials. In the case of polymers a number of techniques are used, which could involve the chemical route and, therefore, not very friendly to the environment. In metals, post-processing is more focused on machining to reduce surface roughness and increase precision, or even for heat treatments aimed at increasing hardness and/or reducing residual stresses generated by the process. Even regarding metals, post-processing is not environmentally friendly, as it requires high levels of electrical energy. It usually generates (recyclable) chips, but it can also use cutting fluids that are normally harmful to the environment.
- **Allows the manufacture of sets of already assembled components:** As mentioned earlier and depending on the designer's expertise the number of individual components can be significantly decreased by grouping them together. Some sets of components that were designed individually and with their own requirements to be assembled later can now be designed and manufactured jointly, either joining components together or even manufacturing components

that are already in the corresponding working position, thereby avoiding work and saving resources.

- **It consists of a series of technologies integrated into the additive manufacturing family:** The diversity of technologies that have been developed around a common nomenclature, additive manufacturing, allows the manufacturer to accurately determine which technology is most convenient depending on the product he wants to manufacture, the required characteristics, and the material that needs to be used. Given that the material normally determines the technology to be used, there are still a number of technologies that allow to select the one that best meets the customer's needs or product requirements. However, despite the diversity of existing technologies, all of them have high environmental sustainability. In fact, resource consumption is much lower and more optimized. In addition, energy is consumed more efficiently. Furthermore, these processes are usually highly automated, fitting perfectly into the Smart Manufacturing philosophy. Since Smart Manufacturing is designed to maximize efficiency, environmental sustainability can also be strengthened in this way.
- **Requires some training in new technologies, but allows relatively quick learning:** In additive manufacturing, human intervention is relatively reduced in manufacturing itself. Essentially, special attention should be given to the component design phase, selection of the technology to be adopted, choice of parameters to be used, and preparation for starting the manufacturing process. This phase requires adequate initial training and some field experience. Given that it is an area largely influenced by technology, operators need to have a high capacity to deal with 3D design software, "slicer" software, and adequate preparation for programming the operating conditions of the production equipment. After that, the process occurs without human intervention until the components are manufactured in the raw. After this phase, human intervention will be necessary, which will need to be specialized, with a view to removing the components from the base where they were built, cutting supports if they exist, and performing the required post-processing, depending on the material, the technology used, the geometry of the components, and how their manufacture has been planned based on the work of the equipment. Thus, if the first phase is strictly technological but requires high attention so that the process safeguards the principles of sustainability, adapting in terms of quality and saving in terms of consumed resources, the last phase depends more on manual dexterity or equipment used, trying to optimize the finishing phase of the components. Given that at this stage, resources that are harmful to the environment may be used, from chemicals to abrasive materials, among others, having an adequate environmental awareness will be a requirement in the training and selection of operators for this final phase of fabrication of additively manufactured components.
- **It leads to savings in the raw material consumed:** As this process deposits raw material precisely where it is needed, except the eventual use of supports, this process, unlike some of the conventional processes, avoids waste of raw material and is therefore considered environmentally sustainable. Furthermore,

as previously mentioned, there are situations in which the infill does not need to be 100%. As it is extremely easy to produce cross-linked structures through this process, this production technique is often used to reduce the weight of components, thus assuring their function and allowing the saving of raw material. It should be noted that it is difficult to reproduce this approach in conventional processes, where the use of material is much more intensive and difficult to obtain .

- **Process usually does not need tools:** One of the main characteristics of different additive manufacturing technologies is that they do not require tools. In most cases, the components are produced on a specific board, from which they will have to be cut at the end of the deposition process. In other cases, mainly regarding biomaterials and photosensitive polymers, components, prostheses, or inserts can be produced from the photosensitive polymer itself or in an aqueous medium in the case of biomaterials. Only in cases of Direct Energy Deposition (DED) or Wire Arc Additive Manufacturing (WAAM) may the process require some type of jig if the repair of a metallic component or composite material is intended. Thus, not needing tools is an added value of this family of manufacturing processes, also showing that it is a highly automated and environmentally sustainable process, as it does not require the consumption of resources in specific tools, as most of the conventional processes require.
- **Process can be used to refurbish, or update worn or broken parts:** Some industrial components, or even other products, wear out and deteriorate during service. This occurs to metallic materials and with plastic and composite materials. Due to the complexity of repair by conventional means, most low-value components were discarded. However, components of greater commercial value or using nobler materials can be reconstituted. Traditional processes require skilled manual labor, which is not always justified. Currently, with 3D scanning technology, it is possible to form a solid exactly with the worn shape it had , simulate the desired shape (probably the one it initially had), conveniently position the worn part inside the equipment (the coordinates will have to be rigorously safeguarded), and rebuild the component using the same material or another that is compatible. Thus , avoiding that many components end their useful life cycle prematurely, prolonging it using additive manufacturing. In these cases, environmental sustainability is safeguarded by extending the life cycle of the product, avoiding recycling or the component being deposited in landfills, and the material used to rebuild it ends up being strictly the essential minimum, thus as the energy consumed by the process. Efficiency is extremely high, thus showing another extremely beneficial aspect for the environment.
- **Overall, it is more energy efficient:** For printing polymers and biomaterials, as well as metallic alloys or composites, energy consumption is essentially restricted to melting, sintering, or curing the material, but in a very localized way. Thus, the energy consumed is greatly reduced. The main loss can be attributed to the use of laser systems, which have a lower energy efficiency than other melting processes. The automation to which this process is subjected means that there is normally only one point where energy is being supplied to the material at any

given time, thus increasing efficiency. As the scanning process allows the heat source to move at high speed, the transmission and loss of heat that is used for the localized melting of the material is greatly reduced, which translates into greater thermal and electrical efficiency.

- **It is more environmentally efficient:** it generates fewer solid effluents, it does not generate liquid effluents, and its emissions essentially depend on how the energy it consumes is generated. In fact, in the manufacture of metallic components, shielding gas can be used, but this gas is inert and harmless to the environment. There are no liquid effluents because cooling in this way is unusual on. The only effluents that result from this process are solid and correspond to the material used in the support structures, and a slight excess of material is needed at the base, so that the cut can be made later. The biggest environmental concern lies in the post-processing processes.
- **Some post-processing techniques can be harmful to the environment:** post-processing, when it does not only include a thermal cycle, normally implies an environmental impact. In polymers, the reduction of surface roughness can be mitigated through chemical treatments, whose products have an environmental impact. When polymeric matrix composites are manufactured, the material that corresponds to the support structures, if used, is more difficult to recycle or reuse because it constitutes a mixture of different materials with quite different properties. The dimensional adjustment of metallic components implies the generation of shavings, which can normally be recycled, implying energy consumption, but not having a very accentuated impact on the environment. However, if cutting fluids are used, they already have a greater impact on the environment. However, if the process is correctly studied, there are alternative solutions that are much more friendly to the environment, such as the use of cryogenic machining or the use of Minimum Quantity Lubricant (MQL). Grinding operations usually lead to the use of abrasive materials, which, in general, are quite harmful for the environment and even for the health of the operators if there is no convenient extraction of the dust generated in the process. As it can be seen, post-processing is the step that raises the most concerns in terms of environmental sustainability. If companies are environmentally responsible, there are alternative techniques that minimize the effects on the environment. Given that companies that usually use this process are at the forefront of technology, they usually have increased concerns about the environment. However, careful analysis will always be necessary in the design phase, so that solutions that impact the environment are not equated.
- **Alliance between complementary technologies (3D scanning and additive manufacturing):** Taking essentially reconstructive medicine into account, the interaction between technologies has constituted a giant step forward in terms of productivity and efficiency in the creation of 100% reliable solutions. While the manufacture of a product involves an idea that is transformed into a virtual solid, which is later converted into a prototype, producing a piece for the reconstitution of a part of the human body follows completely different principles, as each case is different. Modeling a part of a skull or a jaw to replace the original due

to any kind of accident is difficult. 3D scanning assumes a crucial role in this field, as it allows mapping the coordinates of the place where any prosthesis needs to be inserted, allowing this prosthesis to satisfy, in a first iteration, the requirements that are understood as essential for the correct adaptation of the prosthesis. This solution, in addition to constituting another step toward global automation of the process, makes it possible to almost completely guarantee the success of the first and only part to be manufactured to replace the damaged part, without wasting material and time. Here, economic, and environmental sustainability goes hand in hand, also allowing the process to be accelerated, minimizing the suffering of the patient. This solution constitutes a gigantic step in the creation of prostheses, but it can also be extended to old components that need to be replaced. The advantages are perfectly clear in all respects, including the environmental aspect.

- **Logistics:** Logistics seems marginal to this technology, but it is not. Additive manufacturing has brought a huge advantage in logistical terms: production can be carried out locally, if the 3D model is provided by the designer of the components or equipment. This is even more important when people know that the manufacturing process is fast. Given these premises, it no longer makes sense to store critical replacement parts at the customer, or even in intermediate warehouses. If a critical component fails, its replacement can be managed locally and quickly, avoiding the formation of stocks, the possible obsolescence of some products, and the transport of components that are not rarely transported thousands of kilometers between the products and the client. This new approach to spare parts bring a completely different logic to the need for storage and transport over long distances. Manufacturing locally and when there is a real need for the product is something that can be perfectly fulfilled by this technology and that represents clear advantages in terms of environmental and economic sustainability. Effectively, less transport corresponds to less emissions, fulfilling the precepts of environmental sustainability, and stopping producing for stock fulfills one of the principles of the Lean philosophy, which supports economic sustainability, in addition to the immediate saving of resources, which also supports environmental sustainability.

The large series of advantages brought about by additive manufacturing was tested during the COVID-19 pandemic. Due to the flexibility of the technology, the relatively low cost of the FDM technology, and the circulation restrictions imposed by successive confinements, many hospitals were locally supplied with face shields, test swabs, and medical equipment components by local manufacturers and volunteers. It has become possible to manufacture many ventilators based on electronically shared projects and drawings, allowing the skip of the design phase and straight to execution. As no tools were needed, many components could be manufactured locally and simultaneously , multiplying the productive capacity, instead of centralized manufacturing in just a few specialized companies, and reaching the market quickly. In a period when logistics had serious constraints due to the blockage of many supply chains, decentralized production carried out where the products were effectively needed, brought the hope that everyone needed in such a troubled moment for the health of humanity. No country was prepared for such a rapid need and demand for certain products that, until then, had an extremely limited demand. The impact of the pandemic highlighted many of the already known advantages of additive manufacturing, in this case at the service of an extremely noble cause: saving lives.

Given all that has been mentioned above, additive manufacturing is a highly automated process that is in line with the requirements of the Triple Bottom Line: economic, environmental, and social sustainability. The contribution of this process to environmental sustainability was perfectly clear through the advantages that were previously mentioned, considering the main characteristics of the process. Social sustainability is also improved through this process, as operators do not need to perform repetitive and very frequent operations, nor are there clear risks of inducing occupational diseases. In fact, there is some intellectual effort at the beginning of the process, through the design, improvement of the project, and selection of the best operating conditions for the equipment, while the final part, the post-processing, may require some physical effort and present some pollution risks. However, if adequate working conditions are provided, the risks for operators will be minimal. Economic sustainability is the most difficult to sustain In fact, the development of these technologies involved enormous costs, and the companies that invested in their development want a return on that investment, with all legitimacy. Thus, it is possible to verify that some equipment is still extremely expensive, reflecting the investment made in its development. This translates into a greater attribution of this cost to the product, whereby most of the components, mainly metallic and in composite materials, are still not competitive in most industrial sectors with conventional processes. However, there are areas of activity where the development of this technology represented an enormous evolution in qualitative terms, as in the case of the aerospace and aeronautics industries.

Economic sustainability is the most difficult to sustain In fact, the development of these technologies involved enormous costs, and the companies that invested in their development want a return on that investment, with all legitimacy. Thus, it is possible to verify that some equipment is still extremely expensive, reflecting the investment made in its development. This translates into a greater attribution of this cost to the product, whereby most of the components, mainly metallic and in composite materials, are still not competitive in most industrial sectors with conventional processes. However, there are areas of activity where the development of this technology represented an enormous evolution in qualitative terms, as in the case of the aerospace and aeronautics industries. General Electric (GE), realizing the potential of this technology in the manufacture of numerous components for aircraft engines, partnered with companies that developed equipment and manufacturing processes in the area of additive manufacturing, such as Avio Aero in 2013, dedicated to the production of turbine blades for aircraft engines, or Arcam in 2016, which developed and produced additive manufacturing equipment, acquiring them and integrating them into its conglomerate of companies, through GE Aviation and GE Additive. This fact allows to understand that in the last decade, approximately, this manufacturing process was already used for the manufacture of turbine blades for aeronautics, thus contributing to environmental sustainability and energy efficiency. In the field of polymers, the democratization of the use of FDM technology after the original patent expired allowed for an exponential development of the process and a sharp drop in prices, which allowed the creation of new paradigms around this type of manufacturing, such as: (a) hubs were created in which the files with the design of the parts are sent from anywhere in the world, the components are printed by FDM and sent to the customer with very short lead times; (b) the spare parts can be manufactured by the user himself, or by any nearby company, drastically reducing stocks, delivery time, and all the logistics surrounding shipping. Obviously, even in this fact, technology is disruptive: business models no longer follow the standard patterns, to start to be seen in a different way, given the new reality. As the automotive industry deals with lower sales margins, the penetration of additive manufacturing in this sector has registered a positive evolution, but not as fast as could be expected.

In fact, and despite some companies considering that technologies are sufficiently mature to be able to move toward mass production of components, the speed provided by the processes, the cost of equipment, and the cost of raw materials, mainly for metallic alloys and composite materials, it is still necessary to have a convergence of interests between producers and consumers, making processes more expeditious and more economical, making the process sufficiently appealing for a greater number of sectors to adopt this technology. At the same time, the environment would be much better, and the necessary technological leap would be made for additive manufacturing to become the first option in terms of industrial production.

As with all other technologies that break with pre-established paradigms, additive manufacturing will progressively become more economically accessible, and the market will gradually discover all its advantages, gradually adhering to this comprehensive manufacturing process. The possibility of additive manufacturing being integrated into the concept of Industry 4.0, or Smart Manufacturing, will also help in the sustained growth of the application of this technology, which complies with the environmental requirements that are becoming increasingly mandatory for the good of humanity .

6.2 Studies on additive manufacturing sustainability

For all that has been said above, additive manufacturing, in addition to the excellent characteristics it presents as a solution to many of the previously existing limitations in terms of production, also presents numerous advantages in terms of sustainability. Thus, this subject has been the subject of several studies that intend to quantify and dissect the sustainability of these processes, helping decision-makers to have one more positive aspect to consider in their technological and management decisions when launching or reformulating a given product. The environmental aspect is also an increasingly strong selling point, as more and more people become aware of the state of degradation in which the world finds itself, mainly through extreme weather phenomena.

The paradigm shift in the production of goods cannot be achieved with a single click. After the decision-makers have a clear idea about the potential of the process, they should manage the development of new products according to new principles, which requires the corresponding adaptation, research, and training. In fact, in addition to a decision to change the manufacturing paradigm, there is a need to rethink the product for the use of this technology. Effectively, it is not a question of replacing one manufacturing process with another, the product must be studied again in terms of the new potential that additive manufacturing offers. This need for adaptation, which is a time-consuming task, can discourage decision-makers. Change requires a strong conviction, and for that, it is necessary to bear in mind all the previously mentioned advantages.

Javaid et al. [3] deepened the study of the environmental sustainability of the additive manufacturing process, considering the different sectors and applications where it can be used. After a thorough review of literature, the authors dissected the entire ecosystem in which additive manufacturing is inserted, considering aspects related to environmental sustainability. These authors reaffirmed some of the ideas already widely known and previously dissected, namely that the process generates lighter components, consumes less raw material, and is more energetically efficient. In addition, it allows the production of components that ensure a longer useful life cycle for the system in which they are inserted, as they allow the easy and quick replacement of any component that is affected by a catastrophic failure. If the project is properly cared for, it allows the creation of modular systems that can be more easily replaced or updated, which also increases the useful life of a system, avoiding early total replacements. The same authors refer to another critical detail in terms of economic and environmental sustainability: because of more versatile and decentralized manufacturing, supply chains can be much shorter between products and the final consumer, a situation that economically favors the product and minimizes logistical operations, which normally lead to greenhouse gas emissions. Thus, local production, i.e., closer to the final consumer, favors both environmental and economic sustainability, which boosts the business and can attract those who need to make decisions more widely.

In that same study and considering the ecosystem corresponding to the sustainability of the additive manufacturing process, four main aspects were first identified, society, economy, environment, and safety. Around each of these aspects, three factors were identified that can positively or negatively influence the sustainability of the process (Fig. 6.1). Despite the relevance of all the factors mentioned, concerns about post-processing tasks seem not to have been considered, which usually generate solid effluents that are harmful to the environment, such as residues of abrasive materials, which, in most in most cases, do not even have easy recycling. This aspect should also be linked to safety, as the grinding operations of metallic or composite parts carried out by additive manufacturing can release particles that are tiny enough to be inhaled by operators, constituting occupational illnesses when repeated for considerable periods of time. Aspects of facilitating logistics were also not considered in the factors mentioned, but they are undoubtedly positive for both economic and environmental sustainability. The diagram in Fig. 6.1 deserves a deep reflection when thinking in terms of society and in the way that additive manufacturing can help to save and improve the comfort of human lives.

In the conclusions of the work by Javaid et al. [3], the problem of packaging is focused, and a revolution is expected in this market. It is unclear how additive manufacturing can help in the environmental sustainability of the packaging sector,

Figure 6.1. Ecosystem environment for sustainable additive manufacturing [3].

but this is a serious problem that affects the environment, as it is well known to feel that a large part of the waste produced by humanity is related to the packaging of products [4]. If there has been progress in limiting the use of plastics in packaging, normally through regulations that have become progressively more restrictive, there is much to be done in this field. However, due to the still limited production rate of additive manufacturing and the cost of materials involved, this sector does not appear to be close to having any practical environmental benefits from this technology. Only reductions in weight, the number of components, and local production, often in-house, will be able to minimize the environmental impact of the packaging sector.

It is unanimous that additive manufacturing allows for a reduction in the consumption of raw materials and is more energy efficient than conventional manufacturing processes. The generation of solid effluents in the form of garbage, recyclable or not, is also a widely recognized advantage. However, some materials intensively used in processes such as FDM, such as polyamides (PA), acrylonitrile-butadiene styrene (ABS), and polylactic acid (PLA), even though they are recyclable and even biodegradable. In the case of PLA, they still pose problems for the environment and for the health of operators, given that they release Volatile Organic Compounds (VOC) such as ethylbenzene, butanol, cyclohexanone, and styrene [5]. The temperature reached by the heat provided by the equipment to soften these materials so that they can be deposited by extrusion is sufficient to generate the release of those harmful substances. When components are printed in a closed compartment, the accumulation of these VOCs creates a denser concentration of those gasses which are released when the hood of the equipment is removed after the print job has been completed [6]. It should also be noted that in the case of FDM technology, yarn of the same material can lead to different amounts of VOC, depending on the manufacturer, the additives used in its production, and even the diameter of the filament [5]. The VOC emission and ultrafine particle emission rates also vary from material to material. A comparative study conducted between ABS and PLA concluded that ultrafine particle emissions are about a magnitude higher for ABS than for PLA [7]. However, both PLA and ABS are considered ultrafine particle emitters; therefore, their use should not be conducted in closed environments or without adequate air recirculation and very careful particle filtering to minimize the impact on the health of operators. Styrene is the most common VOC [8]. The average diameter of the emitted ultrafine particles varies from material to material. In general, the average diameter of these ultrafine particles is approximately 30 nm, assuming values higher than the average for High-Impact Polystyrene (HIPS), where it is approximately 40 nm, and for glycol-modified polyethylene terephthalate (PETG), which is approximately 60 nm.

It is also well known that 3D printing of polymers releases emissions of organic chemicals, which can lead to cardiovascular and respiratory diseases, asthma outbreaks, and acute hypertension, the latter two of which are mainly caused by inhalation of gasses produced by ABS filament extrusion [9]. A study conducted on animals exposed to the inhalation of gasses from 3D printers in the extrusion of polymeric materials showed significant changes in blood pressure, recording about 2% in resting arteriolar tone [10]. The type and quantity of these emissions depend on the color of the filament, the characteristics of the raw material, the deposition parameters, and even the type of 3D printer used [9,11-12]. As the temperature required to melt or soften the ABS filament is higher than that required for PLA, the number of particles emitted is also much higher because particle emission is largely related to the temperature reached in the extrusion nozzle. If auxiliary heating techniques are used, both build plate and the extrusion nozzle, the emission of these types of gasses can be reduced by approximately 75%, which is much less harmful to the health of operators and the environment [13]. If the 3D printing of polymers through FDM technology causes health problems that can already be considered serious, technologies that use laser for melting/softening polymeric raw materials can even cause high risks of death when humans are exposed to gasses and ultra-fine particles produced by this technology in a more intense way and prolonged time [14]. Studies are still not conclusive regarding the level of gas concentration and exposure time that can cause serious health problems for operators, but it is advised that this type of equipment is not used in environments with little air circulation, namely private homes, libraries, and school laboratories.

Some by-products produced by 3D printing in general are perfectly identified by health organizations as truly harmful to human health. Styrene can be emitted in the form of VOC or ultra-fine particles when HIPS or ABS filaments are used, and it has been identified by the International Agency for Research on Cancer (IARC) as a carcinogenic element. With regard to caprolactam, studies are still ongoing to confirm whether or not this by-product is carcinogenic, with strong possibilities that it is not, and is usually emitted when filaments of PA, plasticized copolyimide thermoplastic elastomer (PCTPE), brick, or wood are extruded [15].

6.3 Additive manufacturing using recycled materials

Additive manufacturing has another aspect that also contributes positively to environmental sustainability, as it allows the use of previously used materials that have since been recycled. The use of recycled materials covers essentially four categories: plastics, metals, ceramics, and composites. Below is a summary of some studies conducted in this area.

The use of plastics/polymers is extremely intense worldwide and represents a serious problem for humanity, as traces of microplastics are already found in marine species and humans. There are still not sufficiently attractive and economical solutions for replacing plastics in the packaging of many products. Not all countries have collection, sorting, and recycling programs for these plastics, and many end up in the sea, posing a serious environmental problem . In countries where there are recycling programs and where the educational level of the population collaborates in this type of action in favor of the environment, there is enough raw material to be transformed into filament and reused again, now in the form of other products. performed through additive manufacturing [16].

PLA is one of the most used materials in additive manufacturing and has also been the subject of several studies that sought to understand which properties are patented by recycled PLA [17-20], comparing the properties patented by the virgin material and by the one that has already been used and was recycled. There are studies focused essentially on improving the properties of recycled PLA [21-22]. However, there are also studies on other polymers widely used in additive manufacturing, such as ABS [23-26], polyethylene high-density polyethylene (HDPE) [27-28], terephthalate (PET) [29,30], polyamides (PA) [31-32], polycarbonate (PC) [33-34], and poly(ethylene-2,5-furandicarboxylate) (PEF) [35]. Despite numerous studies on the mechanical properties of components made with polymeric materials recycled through additive manufacturing, most refer to something that was already known before: with each production cycle, the material loses mechanical properties, with a gradual deterioration of its properties each time it is recycled [23]. There are studies aimed at improving its properties, with a view to compensate for the known loss of properties [22]. The degradation of polymer properties also depends to a large extent on the processing parameters used and the number of times the same polymer is processed. Temperature and processing time are crucial factors for the degradation of mechanical properties: the higher they are, the greater the degradation is usually felt in the mechanical properties of parts that are produced by recycling these polymers. The conditions of use and the additives used also determine the degree of degradation that can be felt in the mechanical properties. Thus, it is common to use virgin plastic for applications with a higher level of responsibility, blends of virgin polymer with recycled polymer for applications with an intermediate degree of responsibility, and only recycled polymer when applications do not require outstanding mechanical properties. Another factor to consider is the ability of the recycled material to adapt to the format required by 3D printing. There are usually no major restrictions in this field, but it is something that needs to be properly analyzed, considering the presentation required for the raw material to be accepted by the additive manufacturing equipment. Aspects such as softening temperature, uniformity of characteristics and properties, among others, need to be properly studied before considering the use of recycled material. However, given the growing demand for this type of manufacturing process, the use of recycled material is a good way out of the environmental problem created by plastics, which, for the most part, are not easily biodegradable and tend to remain in the environment for several dozens of years.

The recyclability of metals is quite good, the degradation of properties is much rarer and can be more easily compensated than in the case of polymers. As additive manufacturing involving metals normally requires more powerful and expensive equipment, the sectors that gave the first impetus to the manufacture of parts by additive manufacturing in metallic alloys were the aerospace and aeronautics industries. This industry is very focused on titanium, nickel, and aluminum alloys. Therefore, it is not surprising that most of the studies conducted on the use of recycled material in additive manufacturing are essentially linked to titanium and nickel alloys, as these are the alloys usually used in propellant and structural components, where high mechanical resistance and low specific weight are required. The use of metallic alloys, as in the case of polymers, can be carried out in the form of powder (Selective Laser Melting (SLM), etc.) or wire (Wire Arc Additive Manufacturing (WAAM)). Its processing is quite demanding, mainly in the case of powder; hence, its cost is quite high. In the case of wire, the process is the same as the Metal Inert Gas (MIG) welding process, so the requirements are a little lower and the cost is also lower. Several studies have focused on the recycling of metal alloys of titanium [36-40], nickel [41-43], aluminum [44-45], stainless steel [46-47], copper [48-49], and magnesium [50]. In addition to metallic alloys, rare earths have also deserved the attention of researchers, with the use of zirconium wire [51] and samarium and neodymium magnetic filaments for the additive manufacturing of magnets [52]. Although metals are relatively easy to recycle through smelting, additive manufacturing provides yet another solution for the possible recycling of metals and alloys. The purity normally required at this stage by many additive manufacturing applications today requires special care in the recycling and treatment of these metal alloys. Furthermore, the forms of presentation of metallic alloys for additive manufacturing are quite specific. However, metal refining and reuse processes are well developed and optimized, making it possible to place them in powder form with the appropriate composition, even in very specific alloys [53-56]. Given that most additive manufacturing technologies for metals work with powder described earlier for the intended composition of the components to be produced, a study was also carried out by Fullenwider et al. [57], who carried out single tracks by additive manufacturing with recycled powders from chips resulting from machining processes and subjected them to a double ball-milling step, stating that the deposited material had greater hardness than the chips that were in the powder used in the realization of this single track. In this way, it appears that recycling in metals is easier to perform and produces better results than in the case of polymers. In metals, waste from conventional processes can be the source of raw materials later consumed by additive manufacturing technologies, which constitutes an added value in terms of environmental sustainability.

Ceramic materials are of particular interest to engineering because of their very peculiar characteristics of hardness and resistance to high temperatures. In addition, the category of ceramic materials presents several solutions from concrete for civil construction, glass for applications where transparency and resistance to high temperatures are required, and the so-called technical ceramics, which can act as electrical insulators or as coatings that can be subjected to high temperatures in ovens. Despite the wide spectrum presented by these materials and due to some versatility in their conventional and non-conventional processing, their recycling and use in additive manufacturing has not yet been decisively explored, with only few studies in this field. Additive manufacturing with ceramic materials can range from being somewhat easier in civil construction, with the deposit of successive layers of concrete, which can house some recycled by-products, to operations that can be daunting tasks to carry out, using directly ceramic powders as raw material, but where multi-step processing can be necessary [58]. Ceramic AM usually uses ceramic powders, while selective laser sintering and binder jet 3D printing are utilized as additive manufacturing technologies [59-60], ceramic slurries comprising ultraviolet (UV) curable photopolymers through stereolithography and digital light processing techniques [61-62], polymeric filaments containing ceramic powders using fused deposition modeling technology [63], ceramic inks/pastes using inkjet printing/direct ink writing technology [64-65], or recycled soda-lime glass printing inks using robocasting technique [66]. Other authors [67-68] studied the use of recycled glass in concrete mixtures to be used in construction carried out through additive manufacturing, showing that the flow properties of this aggregate are better than the usual sand-based concrete, but the mechanical properties do not reach the same level of the usual concrete.

Composite materials have greatly expanded their application spectrum and are also already used with some frequency in additive manufacturing, as their higher mechanical resistance, low weight, longer useful life, and less maintenance make them suitable for more demanding applications. In environmental terms, composites are not the best solution because they always represent an aggregate of materials with significantly different properties, which translates into increased recycling difficulties. As a result of their specificity, they are also not easily reusable in applications other than those for which they were designed. Several studies have also been conducted with a view to integrate composite materials into raw materials that could be used later in additive manufacturing. Wang et al. [69] studied the possibility of integrating milled carbon fiber in the production of recycled PA12 filament. Mechanical tests carried out on samples of the recycled PA12 + 30 wt.% recycled carbon fiber showed interesting improved properties, with tensile strength, tensile modulus, flexural strength, flexural modulus, and impact strength improved by 35, 163, 61, 138, and 23% relative to recycled PA12. Other authors have also studied other approaches based on other materials. Veer et al. [70] reinforced recycled polypropylene (PP) with new fiberglass to produce a 3D printing filament. Tests were conducted with samples printed in different directions, and the results were compared with the values existing in literature for virgin PA12. This study compared the mechanical properties of parts produced by additive manufacturing using glass-fiber reinforced PP filament from two different sources: domestic waste and industrial waste. The study concluded that the degradation of properties is essentially due to the reprocessing of PP and that PP recycled from household waste promotes greater degradation of mechanical properties than PP from industrial waste. The inclusion of metallic particles of elements such as aluminum, silicon, chromium, and even iron in polypropylene (PP)/high-density polyethylene (HDPE) matrices has also been studied to manufacture a filament for use in FDM technology [71]. This composite filament was compared with the recycled PP/HDPE filament without any metallic additive. It was found that 1% Fe–Si–Cr or Fe–Si–Al particles promoted greater thermal stability in the filament, leading to a 37% improvement in yield strength and a 17% improvement in Young's modulus. The improvement in the behavior of products manufactured with composite filament produced with recycled polymers was attributed to a greater interfacial adhesion between metallic nanoparticles and polymers, which benefited the resistance to the generation and progression of cracks. Singh et al. [72] studied the thermomechanical behavior of filaments based on Silicon Carbide (SiC) and aluminum oxide (Al_2O_3) in a reinforced HDPE matrix. However, the study only revealed that an insignificant effect was reported on the thermal and rheological properties of the composite filament, noting that the process is important only in environmental terms, helping to reduce polymers resulting from industrial wastes, and to reduce the cost of the filament.

6.4 Environmental implications of Additive Manufacturing

The use of additive manufacturing has a strong upward trend; therefore, it is necessary to consider how this could have implications for the environment. The diversity of technologies and the immensity of materials that are currently being applied in additive manufacturing brings added challenges to any study that intends to be exhaustive. In order to simplify, it can be said that there are five direct implications of additive manufacturing on the environment:

- **Materials:** From what has been said before, and observing what happens in practice, especially when topological optimization is used, there is a significant decrease in the amount of material consumed, which translates into a clear benefit for the environment: reduction. In addition, it was also clear that the ways in which materials are used in additive manufacturing is also not an impediment to a continuous policy to support recycling, allowing the integration of materials and producing composites with properties of interest to the market.
- **Energy consumption:** The energy issue is extremely complex to analyze because it depends on the material and technology used. When using polymers, the energy consumed per unit volume of material produced in a component form can be less than that in traditional processes, but even this depends on the technology used. In the processing of metals by additive manufacturing, in which high-power beams are used for the localized melting of the metal, the energy consumption per unit volume of processed material is higher, and to this must be added the energy consumption in the preparation of powders or wires that are used as raw material, since the post-processing energy can be equal to or even lower than that required after conventional processes. Both the processing of ceramic and composite materials is even more difficult to assess, as the range of temperatures used is quite wide and varies from material to material [73]. However, one should consider the complexity that is possible to obtain in a single step through additive manufacturing, which, using conventional

technologies, could imply the impossibility of performing this process , or an intricate succession of processes and individual tasks, which would lengthen the lead time and make the comparison much more complex. Some energy can also be saved in assemblies because additive manufacturing allows the aggregation of several components into one as the process deals better with complex shapes.

- **Occupational health concerns:** Additive manufacturing processes do not pose serious health risks to operators. In physical terms, proper care is needed in the handling of construction plates, especially regarding the construction of metal parts. In addition, some post-processing processes, such as the support structures' cut, may pose a threat to the physical integrity of operators, thus, individual safety devices are recommended. The biggest concerns come from the inhalation of VOC and nanoparticles generated during the process, which , due to their concentration and persistence over time, could pose serious risks to operators health [74]. Thus, adequate ventilation and fume extraction in the manufacturing zone are advisable. So far, there have been no other possible risks to health and the environment, beyond those mentioned here.
- **Waste generated:** It could be supposed that additive manufacturing produced an almost null amount of waste due to its own principle of conception: depositing material only where it is necessary. Effectively, the waste created by additive manufacturing is significantly lower than that generated by conventional processes. However, there are three determining factors for the volume of waste created by additive manufacturing: (a) the volume of support structures used in the preparation of the manufacturing process; (b) the volume of debris from the post-processing of the manufactured components; (c) the rate and volume of defective parts generated by the process due to programming errors, operating errors, process parameter selection errors, material defects, etc. Therefore, it is not entirely safe to say that additive manufacturing generates a significantly lower waste volume than conventional processes, although under ideal operating conditions, this is an absolute reality [75-76]. Recycling defective parts manufactured by additive manufacturing is a reality, but polymers lose properties with each new production cycle.
- **Parts lifespan:** Although the useful life of components produced through additive manufacturing depends largely on the user, technological evolution, market competition, and the way the product is idealized and designed, projects involving additive manufacturing tend to focus on components intended to fulfill their function for longer periods of time, which agrees with the intention of reducing the ecological footprint, thus also preserving the environment. On the other hand, weight reduction that is traditional in components produced by additive manufacturing, especially when applied in mobility, will lead to a lower number of emissions during the useful life of the component. Conducting Life Cycle Assessment studies is still premature because of the lack of sufficient data for the studies to present the desired reliability of results.

As described above, the additive manufacturing process does not pose significant environmental concerns. However, there are conditions for additive manufacturing to become even more environmentally friendly. Energy saving has already been sufficiently explored, as models have already been developed for various technologies to minimize energy consumption. The work still to be carried out could involve increasing the efficiency of the equipment that generates the energy used in the melting/softening/sintering of the raw material Exhausting gasses as well as filtering them, are strongly recommended because this is a problem that has been thoroughly studied and identified, and there are solutions that will allow this problem to be mitigated. Owing to the diversity of post-processing tasks, there are still not enough studies to identify and quantify the risks and problems for the environment that are involved with each of these post-processing tasks. This is probably an area where it will be necessary to invest in research, with a view to identifying, quantifying, and mitigating the problems brought about by each task, which also depends on the material being worked on.

References

[1] Chiu, Y. Y., Liao, Y. S. and Hou, C. C. (2003). Automatic fabrication for bridged laminated object manufacturing (LOM) process. Journal of Materials Processing Technology, 140(1–3): 179–184. doi: 10.1016/S0924-0136(03)00710-6.

[2] Silva, F. J. G., Campilho, R. D. S. G., Gouveia, R. M., Pinto, G. and Baptista, A. (2018). A novel approach to optimize the design of parts for additive manufacturing. Procedia Manufacturing, 17: 53–61. doi: 10.1016/j.promfg.2018.10.012.

[3] Javaid, M., Haleem, A., Singh, R. P., Suman, R. and Rab, S. (2021). Role of additive manufacturing applications towards environmental sustainability. Advanced Industrial and Engineering Polymer Research, 4: 312–322. doi: 10.1016/j.aiepr.2021.07.005.

[4] Monteiro, J., Silva, F. J. G., Ramos, S. F., Campilho, R. D. S. G. and Fonseca, A. M. (2019). Eco-design and sustainability in packaging: A survey. Procedia Manufacturing, 38: 1741–1749. doi: 10.1016/j.promfg.2020.01.097.

[5] Wojtyła, S., Klama, P. and Baran, T. (2017). Is 3D printing safe? Analysis of the thermal treatment of thermoplastics: ABS, PLA, PET, and nylon. Journal of Occupational and Environmental Hygiene 14(D): 80–85. doi: 10.1080/15459624.2017.1285489.

[6] Afshar-Mohajer, N., Wu, C.-Y., Ladun, T., Rajon, D. A. and Huang, Y. (2015). Characterization of particulate matters and total VOC emissions from a binder jetting 3D printer. Build Environment, 93: 293–301. doi: 10.1016/j.buildenv.2015.07.013.

[7] Stephens, B., Azimi, P., El Orch, Z. and Ramos, T. (2013). Ultrafine particle emissions from desktop 3D printers. Atmospheric Environment, 79: 334–339. doi: 10.1016/J.ATMOSENV.2013.06.050.

[8] Gu, J., Wensing, M., Uhde, E. and Salthammer, T. (2019). Characterization of particulate and gaseous pollutants emitted duringoperation of a desktop 3D printer. Environment International, 123: 476–485. doi: 10.1016/j.envint.2018.12.014.

[9] Stefaniak, A. B., Johnson, A. R., du Preez, S., Hammond, D. R., Wells, J. R., Ham, J. E., et al. (2018). Insights into emissions and exposures from use of industrial-scale additive manufacturing machines, Safety and Health at Work, 10(2): 229–236. doi: 10.1016/J.SHAW.2018.10.003.

[10] Stefaniak, A. B., LeBouf, R. F., Duling, M. G., Yi, J., Abukabda, A. B., McBride, C. R., et al. (2017). Inhalation exposure to three-dimensional printer emissions stimulates acute hypertension and microvascular dysfunction. Toxicology and Applied Pharmacology, 335: 1–5. doi: 10.1016/j.taap.2017.09.016.

[11] Stefaniak, A. B., LeBouf, R. F., Yi, J., Ham, J., Nurkewicz, T., Schwegler-Berry, D. E., et al. (2017). Characterization of chemicalcontaminants generated by a desktop fused depositionmodeling 3-dimensional Printer. Journal of Occupational and Environmental Hygiene, 14: 540–550. doi: 10.1080/15459624.2017.1302589.

[12] Stefaniak, A. B., Johnson, A. R., du Preez, S., Hammond, D. R., Wells, J. R., Ham, J. E., et al. (2019). Evaluation of emissions and exposures at workplaces using desktop 3-dimensional printers. Journal of Chemical Health and Safety, 26: 19–30. doi: 10.1016/J.JCHAS.2018.11.001.

[13] Deng, Y., Cao, S.-J., Chen, A. and Guo, Y. (2016). The impact of manufacturing parameters on submicron particle emissions from a desktop 3D printer in the perspective of emission reduction. Building and Environment, 104: 311–319, doi: 10.1016/J.BUILDENV.2016.05.021.
[14] Yi, J., LeBouf, R. F., Duling, M. G., Nurkiewicz, T., Chen, B. T., Schwegler-Berry, D., et al. (2016). Emission of particulate matter from a desktop three-dimensional (3D) printer. Journal of Toxicology and Environmental Health, Part A, 79: 453–465. doi: 10.1080/15287394.2016.1166467.
[15] Azimi, P., Zhao, D., Pouzet, C., Crain, N. E. and Stephens, B. (2016). Emissions of ultrafine particles and volatile organic compounds from commercially available desktop three-dimensional printers with multiple filaments. Environmental Science & Technology, 50: 1260–1268. doi: 10.1021/acs.est.5b04983.
[16] Khoo, H. (2019). LCA of plastic waste recovery into recycled materials, energy and fuels in Singapore. Resources, Conservation and Recycling, 145: 67–77. doi: 10.1016/j.resconrec.2019.02.010.
[17] Lanzotti, A., Martorelli, M., Maietta, S., Gerbino, S., Penta, F. and Gloria, A. (2019). A comparison between mechanical properties of specimens 3D printed with virgin and recycled PLA. Procedia CIRP, 79: 143–146. doi: 10.1016/j.procir.2019.02.030.
[18] Zhao, P., Rao, C., Gu, F., Sharmin, N. and Fu, J. (2018). Close-looped recycling of polylactic acid used in 3D printing: an experimental investigation and life cycle assessment. Journal of Cleaner Production, 197: 1046–1055. doi:10.1016/j.jclepro.2018.06.275.
[19] Cruz Sanchez, F.A., Boudaoud, H., Hoppe, S. and Camargo, M. (2017). Polymer recycling in an open-source additive manufacturing context: mechanical issues. Additive Manufacturing, 17: 87-105. doi: 10.1016/j.addma.2017.05.013.
[20] Anderson, I. (2017). Mechanical properties of specimens 3D printed with virgin and recycled polylactic acid, 3D print. Additive Manufacturing, 4: 110–115, doi: 10.1089/3dp.2016.0054.
[21] Paciorek-Sadowska, J., Borowicz, M. and Isbrandt, M. (2019). Newpoly (lactide-urethane-isocyanurate) foams based on bio-polylactide Waste. Polymers (Basel) 11(3): 481. doi: 10.3390/polym11030481.
[22] Zhao, X. G., Hwang, K.-J., Lee, D., Kim, T. and Kim, N. (2018). Enhanced mechanical properties of self-polymerized polydopamine-coated recycled PLA filament used in 3Dprinting. Applied Surface Science, 441: 381–387. doi: 10.1016/j.apsusc.2018.01.257.
[23] Czyzewski, P., Bielínski, M., Sykutera, D., Jurek, M., Gronowski, M., Ryl, L., et al. (2018). Secondary use of ABS co-polymer recyclates for the manufacture of structural elements using the FFF technology. Rapid Prototyping Journal, 24, 1447–1154, doi: 10.1108/RPJ-03-2017-0042.
[24] Mohammed, M. I., Wilson, D., Gomez-Kervin, E., Tang, B. and Wang, J. (2019). Investigation of closed-loop manufacturing with acrylonitrile butadiene styrene over multiple generations using additive manufacturing. ACS Sustainable Chemistry & Engineering, 7: 13955–13969. doi: 10.1021/acssuschemeng.9b02368.
[25] He, H., Zhan, Z., Zhu, Z., Xue, B., Li, J., Chen, M., et al. (2020). Microscopic morphology, rheological behavior, and mechanical properties of polymers: recycled acrylonitrile-butadiene-styrene/polybutylene terephthalate blends. Journal of Applied Polymer Science, 137(4): 48310. doi: 10.1002/app.48310.
[26] Cunico, M. W. M., Kai, D. A., Cavalheiro, P. M. and de Carvalho, J. (2019). Development and characterisation of 3D printing finishing process applying recycled plastic waste. Virtual and Physical Prototyping, 14: 37–52. doi: 10.1080/17452759.2018.1521248.
[27] Singh, A. K., Patil, B., Hoffmann, N., Saltonstall, B., Doddamani, M. and Gupta, N. (2018). Additive manufacturing of syntactic foams: Part 1:development, properties, and recycling potential offilaments. JOM, 70: 303–309. doi: 10.1007/s11837-017-2734-7.
[28] Chong, S., Pan, G.-T., Khalid, M., Yang, T. C.-K., Hung, S.-T. and Huang, C.-M. (2017). Physical characterization and pre-assessment of recycled high-density polyethylene as 3D printing material. Journal of Polymers and the Environment, 25: 136–145, doi: 10.1007/s10924-016-0793-4.
[29] Zander, N. E., Gillan, M., Burckhard, Z. and Gardea, F. (2019). Recycled polypropylene blends as novel 3D printing materials. Additive Manufacturing, 25: 122–130. doi: 10.1016/j.addma.2018.11.009.
[30] Gu, H., Bashir, Z. and Yang, L. (2019). The re-usability of heat-exposed poly(ethylene terephthalate) powder for laser sintering. Additive Manufacturing, 28: 194–204. doi: 10.1016/j.addma.2019.05.004.

[31] Shi, Y., Zhu, W., Yan, C., Yang, J. and Xia, Z. (2018). Preparation and selective laser sintering of a new nylon elastomer powder. Rapid Prototyping Journal, 24: 1026–1033, doi: 10.1108/RPJ-11-2017-0223.

[32] Chen, P., Tang, M., Zhu, W., Yang, L., Wen, S., Yan, C., et al. (2018). Systematical mechanism of polyamide-12 aging and its microstructural evolution during laser sintering. Polymer Testing, 67: 370–379, doi: 10.1016/j.polymertesting.2018.03.035.

[33] Le, V. T. and Paris, H. (2018). A life cycle assessment-based approach for evaluating the influence of total build height and batch size on the environmental performance of electron beam melting. International Journal of Advanced Manufacturing Technology, 98: 275–288. doi: 10.1007/s00170-018-2264-7.

[34] Reich, M. J., Woern, A. L., Tanikella, N. G. and Pearce, J. M. (2019). Mechanical properties and applications of recycled polycarbonate particle material extrusion-based additive manufacturing. Materials, 12(10): 1642. doi: 10.3390/ma12101642.

[35] Kucherov, F. A., Gordeev, E. G., Kashin, A. S. and Ananikov, V. P. (2017). Three-dimensional printing with biomass-derived PEF for carbon-neutral manufacturing. Angewandte Chemie 56: 15931–15935. doi: 10.1002/anie.201708528.

[36] Popov, V. V., Katz-Demyanetz, A., Garkun, A. and Bamberger, M. (2018). The effect of powder recycling on the mechanical properties and microstructure of electron beam melted Ti-6Al-4V specimens. Additive Manufacturing, 22: 834–843. doi: 10.1016/j.addma.2018.06.003.

[37] Li, C., Jiang, C., Peng, M., Li, T., Yang, Z., Liu, Z., et al. (2018). Proinflammatory and osteolysis-inducing effects of 3D printing Ti6Al4V particles: *in vitro* and *in vivo*. RSC Advances, 8: 2229–2239. doi: 10.1039/c7ra12677h.

[38] Gökelma, M., Celik, D., Tazegul, O., Cimenoglu, H. and Friedrich, B. (2018). Characteristics of TI6AL4V powders recycled from turnings via the HDH technique. Metals 8(5): 336. doi: 10.3390/met8050336.

[39] Denti, L., Sola, A., Defanti, S., Sciancalepore, C. and Bondioli, F. (2019). Effect of powder recycling in laser-based powder bed fusion of Ti-6Al-4V. Manufacturing Technology, 19: 190–196. doi: 10.21062/ujep/268.2019/a/1213-2489/mt/19/2/190.

[40] Plaskitt, R., Halfpenny, A. and Hill, M. (2020). Strain controlled fatigue testing of additive manufactured titanium alloy Ti-6Al-4V, ICAF 2019 – Structural Integrity in the Age of Additive Manufacturing, 43–55. doi: 10.1007/978-3-030-21503-3 4.

[41] Sudbrack, C. K., Lerch, B. A., Smith, T. M., Locci, I. E., Ellis, D. L.,Thompson, A. C., et al. (2018). Impact of powder variability on the microstructure and mechanical behavior of selective laser melted alloy. Proceedings of the 9th International Symposium on Superalloy 718 & Derivatives: Energy, Aerospace, and Industrial Applications, 89–113. doi: 10.1007/978-3-319-89480-5 5.

[42] Cordova, L., Campos, M. and Tinga, T. (2019). Revealing the effects of powder reuse for selective laser melting by powder characterization. JOM 71: 1062–1072. doi: 10.1007/s11837-018-3305-2.

[43] Gruber, H., Henriksson, M., Hryha, E. and Nyborg, L. (2019). Effect of powder recycling in electron beam melting on the surface chemistry of alloy 718 powder. Metallurgical and Materials Transactions A: Physical Metallurgy and Materials Science, 50: 4410–4422. doi: 10.1007/s11661-019-05333-7.

[44] Hadadzadeh, A., Baxter, C., Amirkhiz, B. S. and Mohammadi, M. (2018). Strengthening mechanisms in direct metal laser sintered AlSi10Mg: Comparison between virgin and recycled powders. Additive Manufacturing, 23: 108–120. doi: 10.1016/j.addma.2018.07.014.

[45] Ashkenazi, D. (2019). How aluminum changed the world: a metallurgical revolution through technological and cultural perspectives. Technological Forecasting and Social Change, 143: 101–113. doi: 10.1016/j.techfore.2019.03.011.

[46] Saboori, A., Aversa, A., Bosio, F., Bassini, E., Librera, E., De Chirico, M., et al. (2019). An investigation on the effect of powder recycling on the microstructure and mechanical properties of AISI316L produced by directed energy deposition. Materials Science and Engineering: A, 766: 138360. doi: 10.1016/j.msea.2019.138360.

[47] Heiden, M. J., Deibler, L. A., Rodelas, J. M., Koepke, J. R., Tung, D. J., Saiz, D. J., et al. (2019). Evolution of 316L stainless steel feedstock due to laser powder bed fusion process. Additive Manufacturing, 25: 84–103. doi: 10.1016/j.addma.2018.10.019.

[48] Perry, J., Richer, P., Jodoin, B. and Matte, E. (2019). Pin fin array heat sinks by cold spray additive manufacturing: economics of powder recycling. Journal of Thermal Spray Technology, 28: 144–160. doi: 10.1007/s11666-018-0758-3.
[49] Chen, X., Liu, X., Childs, P., Brandon, N. and Wu, B. (2017). A low cost desktop electrochemical metal 3D printer. Advanced Materials Technologies, 2(10): 1700148. doi: 10.1002/admt.201700148.
[50] Salehi, M., Maleksaeedi, S., Farnoush, H., Nai, M. L. S., Meenashisundaram, G. K. and Gupta, M. (2018). An investigation into interaction between magnesium powder and Ar gas: implications for selective laser melting of magnesium. Powder Technology, 333: 252–261. doi: 10.1016/j.powtec.2018.04.026.
[51] Harooni, A., Iravani, M., Khajepour, A., King, J. M., Khalifa, A. and Gerlich, A. P. (2018). Mechanical properties and microstructures in zirconium deposited by injected powder laser additive manufacturing. Additive Manufacturing, 22: 537–547. doi: 10.1016/j.addma.2018.05.037.
[52] Khazdozian, H. A., Manzano, J. S., Gandha, K., Slowing, I. I. and Nlebedim, I. C. (2018). Recycled Sm-Co bonded magnet filaments for 3D printing of magnets. AIP Advances, 8(5): 056722. doi: 10.1063/1.5007669.
[53] Moghimian, P., Poirié, T., Habibnejad-Korayem, M., Zavala, J. A., Kroeger, J., Marion, F. and Larouche, F. (2021). Metal powders in additive manufacturing: A review on reusability and recyclability of common titanium, nickel and aluminum alloys, Additive Manufacturing, 43: 102017. doi: 10.1016/j.addma.2021.102017.
[54] Le. T.-P., Wang, X. and Seita, M. (2022). An optical-based method to estimate the oxygen content in recycled metal powders for additive manufacturing. Additive Manufacturing, 59, Part A, 103127. doi: 10.1016/j.addma.2022.103127.
[55] Cacace, S., Furlan, V., Sorci, R., Semeraro, Q. and Boccadoro, M. (2022). Using recycled material to produce gas-atomized metal powders for additive manufacturing processes. Journal of Cleaner Production, 268, 122218. doi: 10.1016/j.jclepro.2020.122218.
[56] Powell, D., Rennie, A. E. W., Geekie, L. and Burns, N. (2020). Understanding powder degradation in metal additive manufacturing to allow the upcycling of recycled powders. Journal of Cleaner Production, 268: 122077. doi: 10.1016/j.jclepro.2020.122077.
[57] Fullenwider, B., Kiani, P., Schoenung, J. M. and Ma, K. (2019). Two-stage ball milling of recycled machining chips to create an alternative feedstock powder for metal additive manufacturing. Powder Technology, 342: 562–571. doi: 10.1016/j.powtec.2018.10.023.
[58] Chaudhary, R. P., Parameswaran, C., Idrees, M., Rasaki, A. S., Liu, C., Chen, Z. and Colombo, P. (2022). Additive manufacturing of polymer-derived ceramics: Materials, technologies, properties and potential applications. Progress in Materials Science, 128: 100969. doi: 10.1016/j.pmatsci.2022.100969-
[59] Cao, J., Wang, P., Liu, C., Liu, Z., Wang, P., Wu, J., et al. (2022). Research progress on powder-based laser additive manufacturing technology of ceramics, Journal of Inorganic Materials, 37: 241–254. doi: 10.15541/jim20210590.
[60] Gobbin, F., Elsayed, H., Italiano, A., Adrien, J., Colombo, P. and Maire, E. (2021). Large scale additive manufacturing of artificial stone components using binder jetting and their X-ray microtomography investigations. Open Ceramics, 7: 100162. doi: 10.1016/j.oceram.2021.100162.
[61] Su, F., Liu, Y., Zhang, C., Luo, Z., Cao, J., Liu, Z., et al. (2022). Photopolymerization and reaction sintering enabled generative shaping and material-forming of complex ceramic structures with high performance. Additive Manufacturing, 51: 102651. doi: 10.1016/j.addma.2022.102651.
[62] Chen, Z., Li, D. and Zhou, W. (2012). Process parameters appraisal of fabricating ceramic parts based on stereolithography using the Taguchi method, Proceedings of the Institution of Mechanical Engineers. Part B: Journal of Engineering Manufacture, 226(7): 1249–1258. doi: 10.1177/09544054124426.
[63] Sarraf, F., Abbatinali, E., Gorjan, L., Sebastian, T., Colombo, P., Churakov, S.V., et al. (2021). Effect of MgO sintering additive on mullite structures manufactured by fused deposition modeling (FDM) technology. Journal of the European Ceramic Society, 41(13): 6677–6686. doi: 10.1016/j.jeurceramsoc.2021.06.012.
[64] Liu, C., Qiu, Y., Liu, Y., Xu, K., Zhao, N., Lao, C., et al. (2022). Novel 3D grid porous Li4Ti5O12 thick electrodes fabricated by 3D printing for high performance lithium-ion batteries. Journal of Advanced Ceramics, 11(2): 295–307. doi: 10.1007/s40145-021-0533-7.

[65] Qu, P., Xiong, D., Zhu, Z., Gong, Z., Li, Y., Li, Y., et al. (2021). Inkjet printing additively manufactured multilayer SOFCs using high quality ceramic inks for performance enhancement Additive Manufacturing, 48: 102394. doi: 10.1016/j.addma.2021.102394.

[66] de Moraes, E. G., Ferreira, I. M., Teixeira, L. B., Cartapati, L. H., Souza, M. T. and de Oliveira, A. P. N. (2023). Ceramics International, 49(4): 6554–6562. doi: 10.1016/j.ceramint.2022.10.270.

[67] Andrew, G. H. T., Tay, Y. W. I., Annapareddy, A., Li, M. and Tan, M. J. (2018). Effect of recycled glass gradation in 3D cementitious material printing. Proceedings of the International Conference on Progress in Additive Manufacturing, 50–55, doi: 10.25341/D4F59Z.

[68] Ting, G. H. A., Tay, Y. W. D., Qian, Y. and Tan, M. J. (2019). Utilization of recycled glass for 3D concrete printing: rheological and mechanical properties. Journal of Material Cycles and Waste Management, 21: 994–1003. doi: 10.1007/s10163-019-00857-x.

[69] Wang, L., Kiziltas, A., Mielewski, D. F., Lee, E. C. and Gardner, D. J. (2018). Closed-loop recycling of polyamide12 powder from selective laser sintering into sustainable composites. Journal of Cleaner Production, 195: 765–772. doi: 10.1016/j.jclepro.2018.05.235.

[70] Veer, F. A., Setaki, F., Riemslag, A. C. and Sakkas, P. (2017). The strength and ductility of glass fibre reinforced 3D-printed polypropylene. Heron, 62(2): 85–97. doi: N/A.

[71] Pan, G.-T., Chong, S., Tsai, H.-J., Lu, W.-H. and Yang, T. C.-K. (2018). The effects of iron, silicon, chromium, and aluminum additions on the physical and mechanical properties of recycled 3D printing filaments. Advances in Polymer Technology, 37: 1176–1184. doi: 10.1002/adv.21777.

[72] Singh, N., Singh, R. and Ahuja, I. P. S. (2018). Recycling of polymer waste with SiC/Al2O3 reinforcement for rapid tooling applications. Materials Today Communications, 15: 124–127. doi: 10.1016/j.mtcomm.2018.02.008.

[73] Xu, X., Meteyer, S., Perry, N. and Zhao, Y. F. (2015). Energy consumption model of Binder-jetting additive manufacturing processes, International Journal of Production Research, 53(23): 7005–7015, doi: 10.1080/00207543.2014.937013.

[74] Simon, T. R., Aguilera, G. A. and Zhao, F. (2017). Characterization of particle emission from fuse deposition modeling printers, in: ASME 2017 12th International Manufacturing Science and Engineering Conference Collocated with the JSME/ASME 2017 6th International Conference on Materials and Processing, American Society of Mechanical Engineers, 2017 (V002T01A040). doi: 10.1115/MSEC2017-3007.

[75] Song, R. and Telenko, C. (2017). Material and energy loss due to human and machine error in commercial FDM printers. Journal of Cleaner Production, 148: 895–904. doi: 10.1016/j.jclepro.2017.01.171.

[76] Pham, K. and Dotchev, W. (2008). Deterioration of polyamide powder properties in the laser sintering process. Proceedings of the Institution of Mechanical Engineers, Part C: Journal of Mechanical Engineering Science, 222(11): 2163–2176. doi: 10.1243/09544062JME.

Chapter 7

Advanced Coatings as a Greener Production Process

7.1 Advanced and non-advanced coatings: role of technology

Surfaces are, by nature, a more fragile area in solids structure, as they are the place where molecular chains are interrupted, giving rise to places of higher energy, conducive to establishing connections with the surrounding environment. Given this trait, it is common for phenomena to develop on the surface that are not always desirable, such as oxidation and attrition. Sometimes it is even necessary to change the surface to meet only the aesthetic requirements. However, spending large sums of money on a solid where only the surface needs special care to meet the necessary functional requirements is not economically sustainable or advisable. These needs were noted many hundreds of years ago, and man could realize that the problem could be solved using coatings. The first paints and varnishes did not have the current technological refinements; therefore, their protective effect was moderate, with the decorative effect being predominant. The development of the chemical industry allowed considerable gains in terms of adhesion and protection; however, substances that were harmful to the environment and to human beings began to be used. It was then necessary to move from harmful products. Technological development has meant that these coatings have evolved extremely fast. In paints and varnishes, technological evolution is extremely significant, first in terms of quality, and then in terms of preserving the environment and the health of those who face these products, moving from purely synthetic bases to aqueous bases, preserving high levels of quality and greatly benefiting the environment. This evolution shows that there are many environmental problems that can be overcome, if studies are devoted to the subject, until solutions capable of producing the same functions are found, without harming the environment. In this field, science has produced significant results, but there is still a lot to be done, given the current state of the environment worldwide.

Technology has also taken on prominence in this regard. In industrial terms and taking a great part of the 20th century as a reference, technology was mainly linked to the chemical industry, which developed many solutions that met the requirements established by the industry, but with heavy costs for the environment and human health. Chromium plating performed essentially using electroplating processes, has been used for decades without realizing the consequences of hexavalent chromium for human health. The hexavalent chromium compounds are extremely toxic because of their oxidizing power, being carcinogenic as well, especially if airborne and inhaled, but also if in direct contact with it, causing lung cancer, nose cancer, asthma, among other diseases [1,2]. Due to all these problems for health and the environment, the European Union issued regulations to restrict or even eliminate their use through the Restriction of Hazardous Substances Directive [3]. Usually, many of the chemical products used in electroplating processes produce fumes because of the temperature to which they are subjected in the process, fumes that are harmful to human health and the environment. However, these fumes can be properly aspirated and treated, as is the case in countries where regulations are more restrictive. This regulation, based on knowledge of the harmful effects of hexavalent chromium, prevented chromium from continuing to be used in many countries as a way of improving the surface behavior of many components used in the most diverse applications. Moreover, the European Union has legislated to register the use and regulate the application of these chemicals through the Registration, Evaluation, Authorization and Restriction of Chemicals document [3]. Like hexavalent chromium, other toxic chemicals were also used in many manufacturing facilities across the globe. If some were registered and complied with legislation, the low investment and relatively small space that these plants can have facilitates installation in places close to water streams where, often without any scruples, highly toxic chemical products are leaked, creating serious environmental problems, such as the death of countless fish in the first instance, and public health and more general environmental problems later.

In the technological field, electronics and automation played a key role in the development of new deposition processes that did not resort to baths with a chemical composition known to be harmful to the environment and applied electrical power. These processes assume a physical or chemical aspect, respecting the environment in a much more evident way, although not completely exempt from leaving some environmental footprint. In these advanced processes, the coating is carried out inside a dedicated chamber, which is subjected to vacuum, so that the atomic species can circulate in the desired flow, preferably without contamination. In environmental terms, vacuum only interferes with energy consumption, which cannot be ignored. These coating deposition processes are called advanced because they are based on a very sophisticated technological base but using products mostly harmless to the environment. The greatest development was seen in the control and energy generation systems, which allow the performance of the equipment to be completely different from the former technology to the emergence of these systems. This technology began to develop in the middle of the 20th century, but it only started to have some commercial impact between the 70s and 80s of that century, given that there was greater technological development from that time on. After these initial developments, and in view of the interest triggered by the industry [4], successive developments have been registered, which have allowed for the improvement of processes and a significant improvement in the quality of the coatings obtained, considering the diversity of applications that was emerging in the meantime. These new coatings have revolutionized several business areas, namely by the cutting tools in chipping, lenses used in optics, photovoltaic cells, and even jewelry. Thus, it can be said that advanced coatings are essentially based on three vital developments: (1) evolution of vacuum systems; (2) evolution of electronics, mainly in terms of control; (3) evolution of automation and information technology, which now allow more efficient programming of the entire production cycle, and even concerning the preparation of surfaces before the coating process begins to be carried out. While the hardness obtained through conventional coating deposition systems was reduced or moderate, the compounds achieved through advanced techniques continue to reach increasingly higher hardness values. Thus, while conventional coatings were mainly focused on protection against corrosion and decorative effects (apart from hard chrome plating and others), coatings made using new technologies, in addition to these markets, also began to be widely used to sustain wear in surfaces and reduce friction. In environmental terms, in addition these new coatings produced almost no effluents and did not deal with dangerous chemicals, making it possible to create coatings that increase the useful life of many components, and by reducing friction, reduce the energy absorbed, so that the systems become more efficient in terms of operation. Given the regulations developed in the meantime by most developed countries, considering the increase in knowledge about the harm caused to the environment by more conventional coating techniques, these technologies also began to adopt strategies to make these coatings more environmentally friendly, a situation that will be described later.

7.2 Coating functions, applications, and the environment: less harmful emissions and longer component lifespan

Given that the surfaces correspond to a zone where atomic bonds end, which can be incomplete, giving rise to places of higher energy, this zone of solids is always more susceptible to interactions with the environment or with other solids. As the surface is the interface between a solid and the environment that surrounds it, it is often necessary to give it an appearance that is suited to its function, a situation that can also involve the application of coatings, as has already occurred with more conventional coatings. Thus, there are essentially three types of actions that usually tend to interact and degrade surfaces:

Oxidation and corrosion: The environment contains moisture and other elements that can interact with surfaces. Oxygen is most common, as it is in air and water, causing oxidation and corrosion of most metal surfaces. However, some metals have the correct elements to make their surfaces totally or partially immune to the action of oxygen, as is the case with stainless steels, through the self-formation of a Cr_2O_3 barrier on the surface, which prevents oxygen from producing oxidative or corrosive effects. To this end, stainless steel should have a minimum of 12% Cr in its composition, and even under these conditions, it is not 100% immune to oxygen action. Furthermore, when this layer of Cr_2O_3 is not thick enough or is removed by any mechanical action, the surface of the stainless-steel solid is vulnerable at that point. Other elements or compounds commonly used in industries or chemical processes can be even more aggressive than oxygen, as is the case with pure chlorine, which is commonly used in municipal facilities such as swimming pools Corrosion can take different forms, from generalized corrosion, which, as the name implies, affects the entire surface in the same way, to other more specific forms of corrosion, such as crevice corrosion, galvanic corrosion, selective corrosion, and pitting, with the latter being more localized and deeper. It can thus be seen that the in environmental surrounding the surfaces can be aggressive enough to interact and degrade those surfaces, thereby deteriorating their appearance or functionality.

Friction: Sliding contact between surfaces is also a problem that has triggered a new area of investigation called tribology. Friction poses a problem because of the energy it consumes. In a perfect, frictionless system, efficiency could be 100%. However, the energy consumed by friction deteriorates the efficiency value and represents a very significant consumption of energy in global terms, as this phenomenon affects, to a greater or lesser extent, all contacts where there is relative motion from one surface to another. Friction depends on the materials and conditions (like temperature) involved in contact [5,6]. Each pair of materials triggers its own friction phenomena. Considering the number of existing bearings around the world and the fact that these bearings are also affected by friction in their performance, among many other mechanisms, it will be easy to realize the importance that science has given to this problem. In fact, an extremely high amount of energy is consumed in these contacts, and this affects the environment, as energy generation is still not based only on renewable sources, being produced in part using hydrocarbons, which are responsible for much of the greenhouse gas emissions that affect the environment.

Wear: Wear is a very comprehensive occurrence that significantly affects mechanical components, especially those that are animated by relative movement [7,8]. Wear deteriorates components, significantly reducing their lifetime. Wear can take on several mechanisms, the main ones being abrasion, adhesion, erosion, and cavitation, among other more specific cases. The difference in hardness between materials, the metallurgical compatibility between materials in contact, the speed of hard particles that circulate over softer surfaces, and high surface pressures, which promote subsurface phenomena that give rise to the creation of microcracks and detachment of parts of the surface, are some of the causes of wear. Lubrication plays a key role in minimizing some of these cases. Environmentally, this degradation of surfaces reduces the lifetime of components, forcing them to be replaced prematurely, which constitutes a consumption of resources that could be avoided.

The phenomena described above affect different groups of materials differently. Corrosion is an event that typically affects metals and alloys, which have different levels of resistance depending on the product or element with which they come in contact. Science has promoted a selective choice of materials whose application is more advantageous in terms of the function they perform and the environment in which they are inserted, with some typification existing. However, these effects are not limited to metals and alloys. Certain compounds also affect polymers, leading to their rapid degradation through the disintegration of their molecular chains. As with metallic alloys, the composition can be improved through additives, which improve certain properties, thereby preventing very rapid degradation. This phenomenon extends to many other materials, including natural materials such as wood [9]. Even glass, when used in lenses, is subject to scratching by harder materials or with particularly sharp geometry. This case occurs in glasses widely used by the population, where a small distraction can lead to scratches on the surface of the glass's lenses, which becomes uncomfortable in their current use. Very high transparency coatings, usually based on oxides, are commonly used to protect lenses from these scratches. For a relatively small fee, the lenses are provided with a high-hardness coating, which can significantly increase the lifespan of the lenses. These different realities lead to a very high number of applications, which, for the most part, can be typified according to the groups that are presented below and for which some examples are presented:

Wear: The increase in wear resistance is one of the main reasons why there has been a strong pressure on the study and development of coatings using new deposition technologies [10]. It is extremely easy to see that wear exists in everything that is subject to sliding contact, happening mainly between solids (abrasion), but also occurring when liquids carry very hard particles and are thrown against softer solid surfaces (erosion). However, as mentioned earlier, there are other forms of wear, not so common but also cause considerable damage. In certain situations, abrasion is unavoidable [11]. The tools used in machining processes, such as drilling, turning, and milling, are constantly abraded because of the sliding of the cut material over their surface [12]. Furthermore, it should provide the necessary hardness to the cutting edge so that it is efficient in its cutting function. This sector has been an area in which more studies have been conducted on advanced hardness coatings. In fact, the life of a tool can be drastically improved by applying this type of coating [13]. However, the range of applications is extremely wide, even extending to plastic injection molds, stamping tools In addition to the traditional hard coatings used on tools and other mechanical components, coatings with high transparency have also been developed for application to eyeglasses and lenses of other optical devices to avoid scratching [14,15]. These coatings, without affecting the function of the lenses, significantly increase the life of the lenses, which, due to less careful use, could become scratched, which brings significant inconveniences in their function.

Friction: Friction is an extremely common event, desired in some cases to retain relative motion between solid bodies (conveyors and so on), but undesirable in most cases due to the energy it consumes. Thus, several advanced coatings have been developed with a focus essentially on reducing friction [16,17]. However, efforts have also been made to solve this problem through more conventional coatings [18,19], usually targeting different fields of application. Friction affects many cases of contact, but it is particularly important in some manufacturing processes in the field of mechanical engineering. Effectively, many industrial processes use tools that come into contact with the materials to be worked on, either by chips plucking (machining) [20], stamping [21], or even in polymer and metal injection processes [22]. In addition to consuming energy, friction also causes wear and tear due to the surface stresses it generates, which is doubly harmful to the environment: it wastes energy and reduces the useful life of components, forcing them to be repaired or replaced, which implies consumption of resources. It is therefore not unusual that some developed coatings focus on reducing friction and wear at the same time [19,23]. However, in addition to industrial applications that are important in terms of turnover and the number of studies carried out, friction also needs to be considered in terms of the final application. The sliding of snow on roofs and solar panels is another relevant aspect in functional and environmental terms, since the existence of this friction between the surfaces and the snow implies a greater accumulation of snow, which is harmful to the structure where the panels are supported, and conditions the performance of the panels in terms of renewable energy generation. Taking these factors into account, Barker et al. [24] studied the chemical composition of coatings to be applied on solar panels, stating that hydrophilic or hydrophobic coatings promote a decrease in snow accumulation, i.e., promote a greater flow of snow circulation without it being accumulated on the surface of the panels. As many solar panels are placed in areas that are difficult to access, their cleaning is also a problem. If the friction is small, the retention of dust and rubbish will be less, thereby reducing maintenance needs, increasing the performance of these devices, and increasing their useful life. For this purpose, investigations have also been conducted to determine which coatings produce the best effect in terms of self-cleaning of solar panels [25,26].

Reflexion effect: Glazed surfaces, namely solar panels, and glasses, should have specific properties depending on the requirements they are intended to cover. Solar panels should have low reflectivity to achieve better use of the sun's rays that fall on them. Thus, coatings have been developed, such as ITO/SiNx:H, which essentially aim to avoid the reflection of solar rays, thus increasing the efficiency of the panels [27]. On the other hand, solar roof tiles that are used as an alternative way to produce renewable energy from solar incidence, but with a configuration different from that usually presented by solar panels, have an electrical efficiency that is affected by temperature. As the temperature increases in the solar roof tiles, their electrical efficiency decreases, and they may need their own cooling system, which becomes complex and expensive. If solar roof tiles are provided with a reflective coating, the heating caused by sunlight is reduced. This was proved by the work of Alim et al. [28], through the application of reflective coatings, achieving a decrease in temperature of about 11°C and an increase in electrical efficiency of 6.6%. In environmental terms, both anti-reflective and reflective coatings depending on the area and the intended effect, can be valuable aids in increasing efficiency in energy generation.

Thermal barrier: In certain situations, the surface is intended to act as a thermal barrier, preventing the transmission of energy in the form of heat/temperature. Among many other applications, this is useful in cutting tools used in drilling, turning, and milling because the heat generated in the cutting process, if completely transferred to the tool, contributes decisively to its premature degradation. By avoiding heating transfer, the tool will be safeguarded, increasing its lifespan. In this case, several coatings can be applied simultaneously [29], allowing different effects, including thermal barrier. Another very frequent application of coatings made by PVD or plasma spraying as a thermal barrier is in the combustion chambers of aircraft engines. To test the influence of temperature on the useful life of these coatings, Wang et al. [30] studied the influence of three temperature levels on the useful life of thermally grown oxide layer coatings obtained by plasma spraying, concluding that they significantly decreased in terms of a lifespan from 6327 hr to 1642 hr when the temperature was increased from 1000°C to 1200°C. It was verified that each increase in 100°C corresponded to a 50% loss of useful life. This study demonstrates that coatings, despite conveniently performing the function for which they were designed, are subject to intense degradation, a factor that undermines their sustainability. Coatings used as thermal barriers are also subject to other factors that can influence their lifespan, namely corrosion and thermal fatigue, as the corrosion phenomena are often stimulated by exposure to high temperatures. Huang et al. [31] and Xiang et al. [32], among others, investigated the behavior of different types of coatings used as thermal barriers under very aggressive corrosion conditions and tested their resistance to this corrosion. YSZ/TaSi2-MCrAlY coatings demonstrated very high oxidation resistance when exposed to air or water vapor [31]. On the other hand, 5.2Gd2O3–5.6Yb2O3–9.5Y2O3 co-doped ZrO2 (GdYb-YSZ) double-ceramic-layer coatings showed excellent oxidation resistance when exposed to an atmosphere of Na2SO4+V2O5 molten salts, as well as a high resistance to crack propagation, when compared to GdYb-YSZ DCL and single-layer yttria partially stabilized zirconia coatings, both deposited by atmospheric plasma spraying [32]. However, because the cyclic application of temperature leads to thermal fatigue, resistance to cracking has also been studied in coatings commonly used as thermal barriers. Cen et al. [33] developed a framework capable of modeling the thermal barrier coating crack driving force in thermally grown oxide coatings, mainly composed of Al_2O_3, which can block the inward diffusion of oxygen. Yb_2O_3-doped yttria-stabilized zirconia films, usually used as a thermal barrier, provided with different Yb_2O_3 and Y_2O_3 doping contents and obtained using the electron beam physical vapor deposition technique, were tested up to temperatures of 6000 hr at 1150°C in order to analyze its degree of oxidation and resistance to fracture [34]. The authors found that up to around 2000°C there was no delamination of the coating, but there were already cracks that indicated its future detachment. In general, the 8.5YbSZ composition showed worse results in terms of fracture resistance and thermal conductivity than the 4.5Yb4YSZ composition. Blanchard et al. [35] studied the influence of aging caused by temperature on the optical properties and porous structure of coatings usually used as thermal barriers. The deposition temperature of the coating influences the number of pores produced: the higher the deposition temperature, the less pores are formed, which results in less light scattering. Thus, coatings deposited at low temperatures give rise to a greater number of tiny pores, with a diameter equal to or less than 2 μm, which leads to greater light scattering. The application of a heat treatment to these coatings at 1450 K over 1111 hr reduces the spacing between pores, which induces a reduction in light scattering. It can therefore be seen that the range of temperatures normally used by researchers to test the level of degradation of the coatings used as a thermal barrier is between 1000°C and 1200°C, and that in this range of temperatures there is a significant degradation of the coatings in terms of oxidation resistance capacity, resistance to induction and crack propagation, and even light transmission capacity.

Decoration/Style: The possibility of making coatings with different colors, even varying from one color to another, presenting an excellent look and allowing high resistance to abrasion and corrosion, made advanced coatings an extremely attractive option for many decorative applications, even when these parts also had the need for functional requirements. Faucets are included in this class because the golden appearance conferred by advanced coatings, which are more difficult or more expensive to obtain by conventional processes (even using gold), came to be considered as a form of finishing, to the detriment of conventional electroplating coatings, mainly due to environmental issues. Arenas et al. [36] recently studied the application of TiN and ZrN coatings deposited using the PVD technique on tiles for decorative effects. The objective of this work was to study the wear and corrosion behavior of the coating, given the conditions of use to which this type of product is subjected, stating that these coatings can be applied in a domestic environment without significant deterioration, whether due to wear or corrosion. However, the efforts around the study of the applicability of hard coatings for decorative purposes already dated from the last century [37,38], when it was intended to take advantage of the shining golden color normally characteristic of TiN coatings, or the high thermal stability patented by ZrN coating [38], which consumes a much lower energy level than that required for TiN coating, although presenting a different color. In addition, a mixed technique of initial TiN deposition by PVD was used, followed by a very fine gold electrodeposition, giving the coating the desired appearance, but at a lower cost, due to the difference in price between gold and TiN. When this double layer using different depositing processes is used, special care is required due to the possible difference in brightness between the two materials, which could imply an easy identification of the use of these two materials when the top layer of gold wears off, leaving the PVD coating exposed and with a different sheen. For this, it is necessary to use a strong plasma intensity on the substrate to brighten the TiN PVD coating It was also stated that ZrN presents a better resistance to corrosion than TiN, mainly when thinner layers are used, due to the ability of ZrN to react with the surrounding environment, developing an oxidative layer. However, the difference in hue between TiN and ZrN usually conveys these coatings to different types of applications. Environmental concerns also date back to that time, given that TiAl (C, O, N) PVD coatings were developed to replace coatings which were recognized as having very harmful effects on the environment, such as Ti–C:H [38]. This type of coating essentially presents a neutral dark tone. When it is intended that the coating assumes a blue color, the most economical option involves the use of Ti(N,O) coating, which allows obtaining remarkable and clearly defined blue tones. Recently, He et al. [39] studied the effect of adding Fe in the modification of Cr/CrN composite coatings to improve the performance and corrosion resistance when applied to decorative coatings. The coating containing $Cr_{90}Fe_{10}$ (at.%) target/alloy composition showed greater reflectivity within the wavelengths corresponding to visible light and better corrosion resistance than single-phase CrN coatings.

These characteristics are usually attributed to the finer microstructure presented by the newly developed coating containing 10% Fe. Moreover, as the microstructure is finer, the coating is more compact, resulting in less rough surfaces and therefore more suitable for decorative purposes. The color of the coating can also be controlled with the aid of two factors: the Fe content added to the Cr and the N_2 pressure used in the deposition process. The results obtained indicate that the addition of Fe induces better mechanical properties of the deposited layer, reduces the overall cost of the coating, and adjusts the tonality of the coating obtained through the regulation of deposition parameters. The market also requires black coatings, which can also be produced by advanced deposition techniques, normally using C. Gupta et al. [40] studied the deposition of hard black coatings on titanium substrates using ion-implantation technology. They stated that an a:C layer is formed on the top surface in which there is a moderate percentage of sp^3-type bonds, which ensures high hardness, and also forms an intermediate layer of TiC between the top/outer layer of the a:C and the Ti substrate, which results from the implantation of C atoms on the initial surface of Ti (substrate). Despite its very thin thickness, around 100 nm, in addition to having the desired appearance in decorative terms, the coating provides increased resistance to corrosion, reduces friction coefficient by around 25%, and increases hardness by around 72%. It is, therefore, a system that largely benefits the surface characteristics of any component in Ti and is also able to confer the desired look. In the case of decorative coatings, the combination of different processes to obtain a multilayer coating has used different approaches and solutions. In addition to the case of Au deposition by electroplating on a film earlier deposited by PVD [38], Shimpi et al. [41] used two different technologies to obtain the same coating, although both can be classified as PVD techniques. In this case, magnetron sputtering, and reactive arc evaporation techniques were used to produce an inner TiN film and an outer Au alloy film. For the TiN coating, the magnetron sputtering process was used, and then the reactive arc evaporation process was used to deposit the Au alloy. This set of processes complemented by cleaning and passivation processes, has made it possible to obtain excellent quality coatings that can be applied to watch straps. The automotive industry has substantially increased the application of advanced coatings for tribological applications, to the detriment of conventional coatings, which have induced serious environmental problems. Ferreira et al. [42] studied the morphology and wear behavior of CrN coatings deposited by cathodic arc deposition with a view to their application in components of combustion engines, stating that the depositions carried out using lower nitrogen flow and higher substrate bias voltage came from denser coatings and with higher hardness, which have a better wear action and lower friction coefficient, compared to coatings also made of CrN deposited by the same process, but with a columnar structure. Despite the deposition of advanced coatings on polymers representing an even greater challenge due to the softening and melting temperatures of most polymers of interest to the automotive industry, Navinsek et al. [43] pointed to advanced coatings as an excellent solution in environmental terms to replace conventional coatings carried out by electroplating, but essentially aimed at metals. However, PVD coatings have already been applied as an anti-reflective element in car optics, anti-scratch coatings, and coatings to facilitate the cleaning of certain interior and exterior surfaces of motor vehicles. In 2003,

Kuhr et al. [44] overcame a barrier that until then had been felt in terms of adhesion of coatings performed by CVD on polymeric substrates through the development of interlayers capable of adhering more effectively to the substrate and allowing subsequent deposition by plasma impulse chemical vapor deposition of coatings with the most suitable thickness and composition for the applications intended by the automotive industry, being essentially SiO_2 based layer system deposited on polymethyl methacrylate (PMMA) substrates.

Biomedical applications and biocompatible coatings: The evolution of some areas of medicine has also created the need to develop coatings capable of interacting with the human body and its fluids, namely, improving the surface properties of prostheses. However, in this case, a thorough study of the biocompatibility of the coatings is necessary to ensure that other pathologies are not created for the patient. There have also been numerous studies conducted in recent years on this matter, using different types of coatings, but almost always resorting to PVD technology to produce these coatings. Zarka et al. [45] studied the application of a complex $Ti_{3.6}Nb_{1.0}Ta_{0.2}Zr_{0.2}$ film on anodized and non-anodized aluminum substrates, dissecting the differences in behavior after the coating process and in the expected work environment. After deposition, it was stated that the addition of the aluminum used as substrate induces a significant decrease in the grain size, improves the adhesion of the coating to the substrate, and positively influences the cracking mechanisms of the coating in service. Thus, it can be observed that the aluminum anodizing treatment is favorable for the subsequent deposition of PVD coatings for biomedical purposes. In this field, the coatings also showed a hydrophilic action and adequate *in vitro* resistance when in contact with body fluids, thus validating their potential use in short-term biomedical applications. A coating identical to the previous one (TiNbTaZr) was also deposited by PVD and tested on pure Mg alloys and AZ31 alloy, with the aim of its use in biomedical applications [46]. The objective of this study was to characterize the corrosion resistance of the coating in contact with body fluids and analyze its wettability. The coating had a dense columnar microstructure with a grain size between 20 nm and 40 nm, and the hydrophilicity of the Ti-based PVD coating was better than that exhibited by the uncoated Mg. However, in more demanding tests of corrosion resistance to simulate its working environment in the human body, there was the formation of some products due to a corrosion event a situation that deserves due care. In addition, in the field of prostheses, in response to the requests imposed on them in service, advanced coatings have played an important role. In fact, prostheses are subject to strong wear and tear and the action of body fluids, which could induce corrosion in these prostheses, which are generally metallic, although they are produced in appropriate alloys to permanently contact the human body. Staszuk et al. [47] studied the applicability of TiN/ZnO coatings produced by combining two different coating technologies, PVD and ALD (Atomic Layer Deposition), on AISI 316L stainless steel substrates to increase corrosion resistance and to analyze the degree of adhesion of TiN/ZnO coatings to those substrates. The study concluded that, in terms of electrochemical properties, when the substrate of AISI 316L coated with TiN/ZnO is exposed to Ringer's solution, it presents an increased resistance to corrosion and an excellent resistance to abrasion. Thus, allergies to which the patient could be exposed as a result of the release of small nickel debris from the corrosion process of the prosthesis during service can be avoided. Hassan et al. [48] studied the same AISI 316L stainless steel substrate but now provided nanometric and micrometric graphite coatings using Physical Vapor Deposition (PVD) to improve hemocompatibility, antithrombosis, and anti-leaching properties. The focus of the study of nano—and micrometric films was to analyze the structure and roughness, optimizing the deposition parameters for the situation that is understood to be more favorable to the application in question. Tests were then carried out to identify its corrosion potential using atomic absorption spectroscopy to detect coating particles released in service by simulating common conditions of use. The existence of a direct relationship between coating thickness and properties such as mechanical strength and smoothness was confirmed. It was also found that the anticoagulation properties improved, and a decrease in platelets adhered to the coated surface was observed. Accelerated aging tests showed that no particles released from the coating were observed, and there was also no redox reaction. Thus, it can be said that graphite coatings significantly improve the corrosion resistance of AISI 316L stainless steels and do not induce the release of toxic particles into the human body, which could lead to complications for the patient 's health.

Through the description of some studies and applications mentioned above, the study and application of advanced coatings in the most diverse areas of business and our daily lives has been a reality for over 40 yr. However, due to the high initial costs of the technology and the development it has known, only in the last two decades has there been a more intensive application of this type of coating. First, the focus was centered on increasing the useful life of many components, then moving on to the area of application in countless parts used in our daily lives, and only in the last 20 yr has a greater environmental concern emerged, which tends to become more pronounced due to market pressure and legislation that governments have launched in the sense of restricting the most polluting conventional coatings.

Electronics and automation have played a leading role in the development of reactors and devices linked to the production of advanced coatings. Sensing and vacuum systems have also seen a significant evolution in the last two decades, which, together with the studies of the deposition processes carried out in the meantime, have allowed an evolution of coatings from simple titanium nitride to extremely complex coatings, both in monolayer and multilayer, and to coatings with evolutionary properties from the substrate to the surface, which have revolutionized the market and become increasingly common. This has helped to reduce their cost and improve their properties through a deeper knowledge of how the manufacturing processes in each technology influence the final properties of coatings.

7.3 Issues regarding advanced coatings use

As mentioned earlier, coatings made using advanced techniques have deposition principles completely different from traditional processes. One of the main factors in this change is its realization in vacuum, with the aim of eliminating contamination. However, these vacuum systems, and electronics, limited the faster evolution of these systems. The need for a vacuum largely limited the size of the parts to be coated, a limitation that was dissipated over time with the introduction of reactors with ever greater capacity and equipped with more powerful vacuum systems, which drastically reduced the cycle time without detracting from the quality of the coating process. Thus, the main advanced surface coating processes are based on vacuum systems, which guarantee the necessary purity of the deposited materials. In addition, they help in substrate cleaning operations, contributing to an increase in the adhesion of coatings to the substrate.

Another factor to consider in these coatings is that their thickness is usually quite thin. Effectively, this thickness can vary from a few tens of nanometers to just over ten micrometers, and even though this thickness is so small, it can contain numerous layers, with a view to preventing the progression of cracks that may be generated on the surface. Sometimes, these layers also have different properties, where each layer has a specific function in fulfilling a certain universal task. In fact, advanced coating technologies allow various coating architectures through the combination of different layers, which significantly improves their aptitude for the functions they need to fulfill. In this way, the designer can select the most convenient architecture depending on the service that the coating will perform, the conditions to which it will be exposed, thus improving its performance and useful life. Obviously, this possibility has also widened the potential use of these coatings, which have been seen as a credible option in an increasing number of cases. The following figure schematically shows the different plannings that can be equated when considering the possibility of using an advanced coating on a given surface (Fig. 7.1).

The need to meet the diversity of applications and specific market requirements has led to all these plannings to exist. The single layer is the easiest to perform and is therefore probably the most economical solution. It has very similar properties from the interface with the substrate to its outer surface, so it is relatively easy to characterize [50,51]. Its complexity is minimal and the results obtained are relatively predictable. The double layer is widely used because it has added advantages over the single layer: the inner layer can have a completely different function from the outer layer, namely, facilitating the adhesion of the outer layer to the substrate, among many others [52]. It is specially used when there is total or partial incompatibility between the required outer layer and the substrate, or when the inner layer has a specific function, such as serving as a thermal barrier, which is not fully compatible with the function that is required in the outer surface [53]. However, this solution creates an intermediate interface that could limit some intended characteristics of the coating. When this issue arises, it becomes necessary to start the deposition with one material and progressively evolve to another. This type of coating is called gradient coating because it starts with one composition and ends with a very different one, but

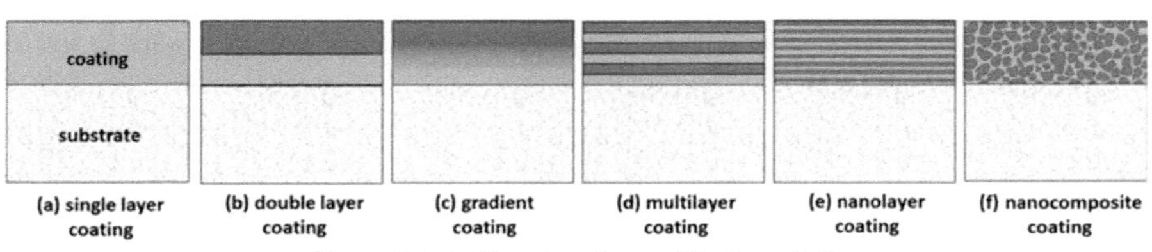

Figure 7.1. Different coating architectures [49].

evolving gradually between the first and the second [54]. It demands programming very specific to the reactors where they are manufactured, but it covers some needs that other types of coating cannot fulfill. The proper evolution of the material and structure from the inside to the outside requires a complex process of selecting materials/gasses and manufacturing parameters. Multilayer planning could imply the alternating deposition of only two types of coating, normally with a well-defined period, which is creased by the thickness of the layers. The thickness of the layers depends on the deposition parameters and rotation of the samples into the reactor [7]. However, it is also often used with more than two different types of film, presenting specific properties, and meeting the needs of the application. It is a type of planning usually found in the coating of cutting tools for machining processes. Films made of nanolayers are usually designed in this way to prevent the easy progression of cracks through these coatings [55]. Each of the many existing interfaces requires greater energy to continue crack progression after it has crossed the previous nanolayer, a situation that makes crack progression more difficult and may even contribute to its dispersion or stoppage. This is exactly the intended effect in these cases, where it is foreseen that cracks will be generated in the coating with a strong probability of being catastrophic. Thus, the planning of the coating intends to counteract the effects that could cause its failure in service. Nanocomposite coatings have been designed for the same purpose as nanostructured coatings, although the structure in this case is less organized [56]. In fact, the progression of cracks along these coatings can be difficult, but there are no perfectly defined interfaces, as in the case of nanostructured coatings. Composite coatings also result from the need to physically join materials that would be impossible to join in alloy form, due to lack of miscibility in the liquid phase, or where it is simpler in terms of the process to co-deposit material from different targets [57]. There are also situations in which multilayers were used together with composite coatings to achieve some properties that are impossible to reach with another type of coating architecture [58].

One of the major problems faced by surface engineering and process engineering is the generation of residual stresses in coatings. In fact, advanced deposition processes generally involve the use of relatively high temperatures. In certain processes, the temperature usually used is even higher, which can cause perfectly expected physical effects, which are then at the beginning of the stress generation [59]. Effectively, the substrates are usually heated before the deposition process with the aim of facilitating the adhesion of the deposited coating material to the substrate. However, depending on the physical/thermal properties of the substrate material, this heating normally generates an expansion of the substrate before the deposition process itself. When the deposition takes place, the substrate is in the expanded state, and the coating acts in the exact measure of its expanded state. However, when the deposition process is over, the substrate is brought back to room temperature, contracting appreciably as much as it had initially expanded. This phenomenon means that the coating must follow the contraction of the substrate, which induces very high compressive stresses in the coating, which can exceed 3 GPa [59]. Normally, the greater the thermal expansion presented by the substrate material and the greater the temperature developed in the process, the greater the level of stress installed in the coatings, which can give rise to several extremely serious events, which can lead to delamination of the coating when the substrate/coating set approaches room temperature. This phenomenon of delamination also depends on the adhesion strength of the coating to the substrate, stiffness, and both elastic and plastic deformation capacity of the coating. To avoid this effect, a commonly used solution is the use of interlayers, which accommodate part of the stress generated at the substrate/coating interface. However, this solution still needs to accommodate the stresses generated, although the stresses can be spread over the interfaces, interlayer, and outer coating. Although the stress is usually compressive because of the occurrence of substrate shrinkage during cooling, this factor can be positive if the stresses to which the coating is subjected under service are of the opposite sign. This effect can even be used in certain circumstances to improve coating performance. To minimize the effect of stresses caused by the deposition thermal cycle, Tu et al. [60] developed a TiN/TiSiN multilayer coating in the form of a gradient by varying the thickness of the different layers used in the coating. It was stated that the thick layers at the base and top of the coating not only improve the mechanical properties of the coating but also contribute to an increase in the level of residual stresses generated by the thermal cycle. However, when the hard layers are thinner and more ductile and thicker layers are intercalated, the level of installed stresses tends to decrease. Zhu et al. [61] also reported situations in which the $V_{0.43}Al_{0.57}N$ coatings reached extremely high stresses, i.e., 3.8 GPa. Considering the various deposition conditions used, it was stated that increasing the bias in the substrate, even with relatively low deposition temperatures (< 200°C), increases the densification of the coating, which translates into an increase in hardness and the level of residual stresses installed in the coating. Thus, it is possible to state that the level of residual stresses is not only affected by the thermal cycle but also by the bias used, a parameter that normally has a direct influence on the compaction of atoms during the deposition process, mainly in the PVD process.

Another aspect that is of vital importance for the correct functioning of coatings is adhesion to the substrate. This adhesion depends on several critical factors: (a) the metallurgical affinity between the substrate and the first deposited layer; (b) possible differences between the physical properties of materials (coefficient of thermal expansion, for example); (c) cleaning the substrate; (d) the thickness of the deposited film; (e) the parameters used in the coating synthesis or deposition process. If the metallurgical affinity is not adequate, the bonding between the substrate surface atoms and the first layers of coating atoms will not develop adequate linkage, and coating adhesion will be poor. Some physical properties become vital for adhesion, such as the coefficient of thermal expansion. Given that the coating process is usually carried out at medium or elevated temperatures, there is expansion of the substrate before and during the deposition process. In the cooling stage, there is shrinkage of the substrate and the coating. If the shrinkage is significantly different between the substrate and coating, stresses that are sufficiently high are generated in the interface, which can compromise adhesion or create the necessary conditions for a possible failure in service. The surfaces are exposed to dirt and contamination, a situation that is very harmful to advanced deposition processes because the coatings are extremely thin and are generated through successively deposited atomic layers. If the interface is not properly cleaned, the linkage between the first layers of synthesized or deposited atoms and molecules will not form the necessary linkage with the substrate, impairing adhesion. In particular, when using the PVD deposition process, making films thicker than 10 mm can generate stresses that induce a significant drop in the adhesion of the coating to the substrate. When the CVD process is used, it is possible to obtain slightly thicker films without significant adhesion problems. Other advanced processes, such as thermal spraying, are more suitable for making thicker coatings without detecting significant adhesion problems, although cleaning and temperature are also critical for adhesion. Parameter optimization is essential for any deposition process, both conventional and advanced. Each deposition technique is governed by a set of parameters, which include temperature, the potential difference between the substrate and the source of material for deposition, the power used, and the vacuum level Simulation techniques for the advanced deposition of coatings are not yet sufficiently developed because of the diversity of parameters, so optimization is normally achieved through experimentation. The optimization of parameters to achieve certain specific characteristics of coatings has led to a large research effort.

7.4 Green coating deposition techniques

Advanced coating production processes, despite not being restricted to vacuum techniques, are essentially based on two broad categories: PVD and CVD. However, each of these categories is further subdivided into several other categories, which always have characteristics that can differentiate them when it is necessary to fulfill certain requirements or obtain certain specific properties. Given their importance, both in the industrial and scientific contexts, these will be the processes to be presented in a further developed way.

7.4.1 Physical Vapor Deposition (PVD)

The PVD process, as the name implies, is based on the transfer of physical species from a solid source to a substrate, which creates the vacuum and temperature conditions for these species to transfer from a source to a substrate. For the deposition to occur under the intended conditions, it is first necessary to generate a low pressure inside the reactor capable of facilitating the transfer of these species and their condensation on the surface of the substrate. In Fig. 7.2, it is possible to observe the most common PVD techniques, although there are other derivatives of these, with small differences, which have given rise to various designations, although these variants are still very specific and without a significant expression in terms of the volume of published studies or frequent use in industry. PVD techniques are divided into two main principles: sputtering and evaporation. Even within these two families, there are different approaches, as illustrated in Fig. 7.2. In sputtering, ionization can be induced in several ways, with the most common being magnetron. Even within magnetron sputtering, there are still several variants, as illustrated in Fig. 7.2. The Radio Frequency (RF) variant is mainly used for polymers because they do not exhibit electrical conductivity [62] but can be used in electrically conductive materials [63]. The Direct Current (DC) variant is one of the most used for depositing coatings on metallic or composite substrates [64]. More recently, the high-power impulse magnetron sputtering (HiPIMS) variant was introduced, which usually

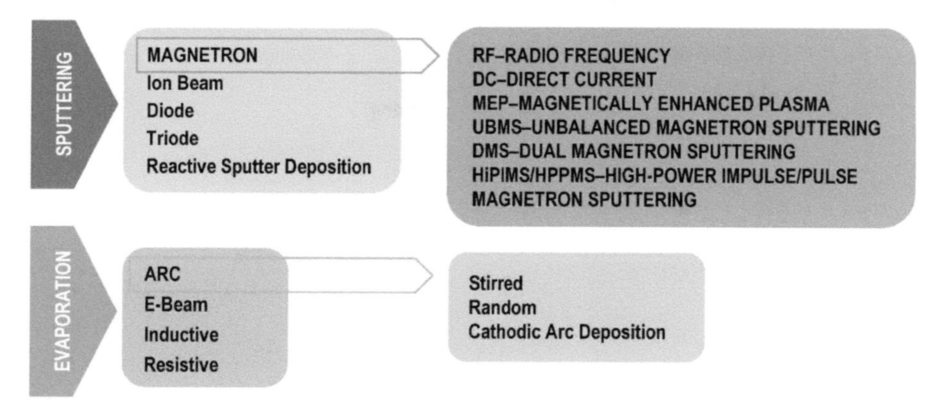

Figure 7.2. Different PVD techniques [4].

promotes a higher deposition rate and better adhesion of the coating to the substrate because it uses very high levels of energy for very short periods of time [65]. This technology promotes much higher levels of ionization of the sputtered metals and a very high rate of dissociation of gas molecules used in the process, which translates into the effects, which are of vital importance for the proper functioning of the coatings: greater productivity and better adhesion properties to the substrate [66]. The high density used in the process results in denser coatings with lower defect rates, which translates into better mechanical properties, such as better wear resistance. The sputtering process consists of projecting atoms of a gas against what is called a target, which contains the material that will be used in the coating. As the magnetic field is created, and the high voltage applied, the gas atoms are violently projected against this target, tearing atoms from it. Atomic particles are pulled from the target, and due to the low-pressure atmosphere into the reactor and acceleration induced by the applied bias, they collide with the substrate surface, initiating the formation of successive atomic layers that constitute the coating layer [4]. The gasses used to extract the atoms from the substrate are usually inert, such as nitrogen and argon, but this depends on the desired coating. Mastering the parameters allows structures to be obtained as planned, i.e., with the desired properties. The roughness level is normally quite low, and the coatings are quite uniform. Given that the gasses used are harmless to the environment and that the process does not generate solid or liquid effluents, this process is considered green.

The evaporation technique is quite different from the sputtering technique, as it essentially promotes the evaporation of the material to be deposited as a coating, creating the necessary conditions within the reactor for the evaporated particles to be deposited on the surface of the substrate [67]. The different variants are essentially associated with how the cladding material is heated until evaporation, the most common technique being the arc [68], as shown in Fig. 7.2, and there are still some variants within the arc technique itself. However in the last decade, evaporation using electron beam has been explored more, as expressed by a greater number of scientific publications in this field [69]. Initially, this technology was essentially used for thicker and rougher coatings, given that the release of material particles from the coating is not uniform and leads to the deposition of particles with somewhat

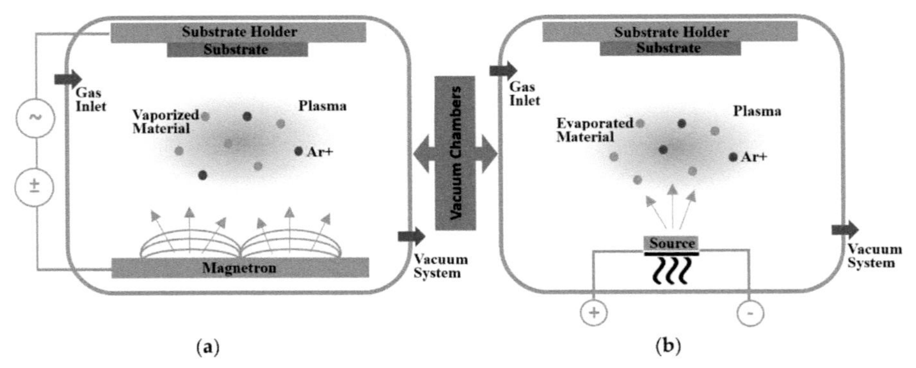

Figure 7.3. Schematic principle of (a) sputtering and (b) evaporation PVD techniques [4].

different sizes, with the main advantage being the high deposition rate. Currently, the evaporation process is much better controlled, allowing to obtain coatings with a quality very similar to that of the sputtering technology [70].

Figure 7.3 schematically illustrates the operating principle of PVD techniques using the sputtering (a) and evaporation (b) principles [4].

7.4.2 Chemical vapor deposition (CVD)

As the name implies, the CVD technique does not resort to a solid source to remove particles from the same source and then project them onto the substrate. Instead, it introduces a mixture of gasses into a chamber subject to low pressure, creating the necessary atomic excitation and reaction conditions to condense these gasses on the substrate in a solid form, forming the coating through the synthesis of these gasses. For the synthesis and condensation of the gasses to occur as intended, it is necessary to promote reactions in the gasses introduced into the reactor, which can be done in different ways, which is exactly what characterizes each of the technologies. Although there is a wide range of processes claimed by several researchers, the two processes that have garnered the most interest from the industry as well as from the scientific community are activation through microwaves [53] and through hot filament (Fig. 7.4) [71]. The principle is quite simple: a previously defined set of gasses capable of generating the intended chemical composition for the coating, is introduced into a reactor subjected to low pressure (vacuum). The substrate where the coating is to be applied is heated. The gasses introduced into the reactor are activated by microwaves or high temperature, causing them to react chemically and precipitate on the substrate, condensing. The temperatures used in this process are usually higher than those used in the PVD process, initially limiting its application to certain substrates. Given that this limitation was identified very early on, numerous efforts were made by the scientific community to reduce the temperature necessary for the decomposition and reaction of the gasses inside the reactor, as this temperature reduction allows for the broadening of the range of substrates that can be coated using this technique and reduces stresses in the coatings deposited [72-74].

CVD coatings are conditioned to the composition of the existing gasses and the chemical reactions that can occur. As in the PVD process, the deposition conditions

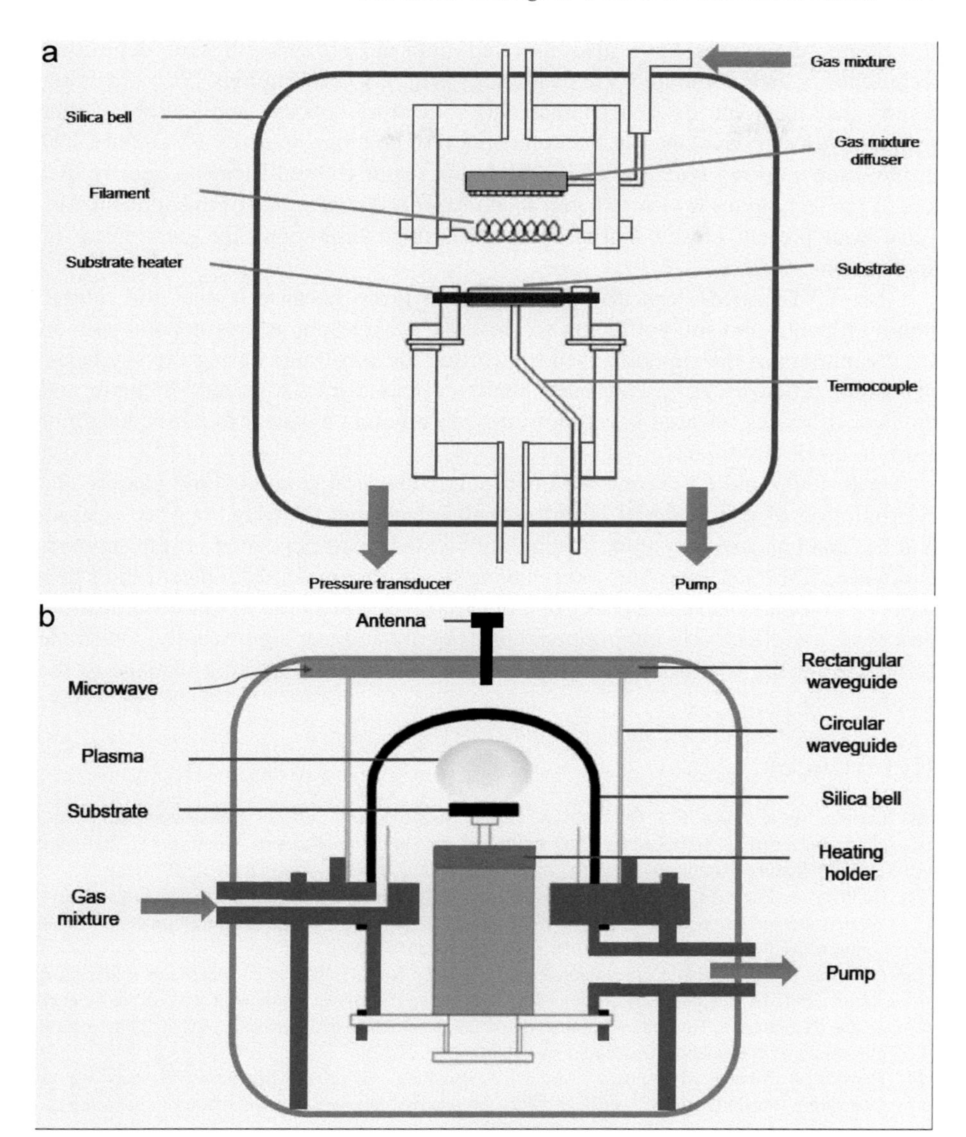

Figure 7.4. Schematic principle of (a) microwave and (b) hot-filament CVD techniques.

can vary during deposition through the introduction or removal of certain gasses causing the coating composition to vary throughout its synthesis. The CVD process is also of special interest because it allows the synthesis of diamond, which, due to its properties of extreme hardness and resistance to wear, assume relevance in the area of cutting tools [71], mainly in the cutting of nonferrous materials [75]. The CVD and PVD processes can be used together, whichever joint use is more favorable to the intended final characteristics. This situation has been explored by tool manufacturers, where a combination of these technologies has been used.

Some researchers have also invested time in exploring hybrid deposition techniques, which combine PVD technology with CVD technology [76], obtaining results that meet the desired characteristics in more specific applications. Other works have been developed in the comparison of coatings made by PVD and CVD in the same working conditions [77,78], but it seems to be difficult to declare that one of the techniques is clearly better than the other for each one of the applications, since both present clear benefits and well-defined limitations for each group of applications studied.

The CVD process can also be considered green because it does not release liquid effluents, and solid effluents are practically irrelevant as they depend only on the cleanliness of the supports used to position the substrates during the synthesis/deposition process. The gaseous combinations released are not particularly toxic, and the flow of gasses released is very low, not representing a danger to public health or harm to the environment.

Both CVD and PVD processes consume considerable energy and gasses. The consumption of raw material is quite small, given that the vast majority is used, and the coatings are very thin, leaving only a small part deposited on the supports and walls of the reactors. Thus, and having conventional deposition processes as a basis for comparison, such as electroplating, it can be stated that advanced deposition processes are effectively environmentally friendly and can significantly contribute to increasing the useful life of many components, without causing damage to the environment.

References

[1] Ferreira, A. A., Silva, F. J. G., Pinto, A. G. and Sousa, V. F. C. (2021). Characterization of thin chromium coatings produced by PVD sputtering for optical applications. Coatings, 11: 215. doi: 10.3390/coatings11020215.

[2] Baptista, A. Pinto, G., Silva, F. J. G., Ferreira, A. A., Pinto, A. G. and Sousa, V. F. C. (2021). Wear characterization of chromium PVD coatings on polymeric substrate for automotive optical components. Coatings, 11: 555. doi: 10.3390/coatings11050555.

[3] European Union. (2014). Commission Regulation (EU) No 301/2014 of 25 March 2014, amending Annex XVII to Regulation (EC) No 1907/2006 of the European Parliament and of the Council on the Registration, Evaluation, Authorisation and Restriction of Chemicals (REACH) as regards chromium VI compounds, (Accessed on January 30th, 2023).

[4] Baptista, A. Silva, F., Porteiro, J., Míguez, J. and Pinto, G. (2018). Sputtering Physical Vapour Deposition (PVD) coatings: A critical review on process improvement and market trend demands. Coatings, 8: 402. doi:10.3390/coatings8110402.

[5] Fernandes, L., Silva, F. J. G., Andrade, M. F., Alexandre, R., Baptista, A. P. M. and Rodrigues, C. (2017). Increasing the stamping tools lifespan by using Mo and B_4C PVD coatings, Surface & Coatings Technology, 325: 107–119. doi: 10.1016/j.surfcoat.2017.06.043.

[6] Nunes, V. Silva, F. J. G., Andrade, M. F., Alexandre, R. and Baptista, A. P. M. (2017). Increasing the lifespan of high-pressure die cast molds subjected to severe wear. Surface & Coatings Technology, 332: 319–331. doi: 10.1016/j.surfcoat.2017.05.098.

[7] Silva, F. J. G., Casais, R. C. B., Baptista, A. P. M., Marques, M. J., Sousa, V. M. C. and Alexandre, R. (2022). Comparative study of the wear behavior of B4C Monolayered and CrN/CrCN/DLC multilayered physical vapor deposition coatings under high contact loads: An experimental analysis. Journal of Tribology, 144(3): 031701. doi: 10.1115/1.4051226.

[8] Martinho, R. P., Silva, F. J. G. and Baptista, A. P. M. (2008). Cutting forces and wear analysis of Si3N4 diamond coated tools in high-speed machining. Vacuum, 82(12): 1415–1420. doi: 10.1016/j.vacuum.2008.03.065.

[9] Costa, R. D. F. S., Barbosa, M. L. S., Silva, F. J. G., Sousa, S. R., Pinto, A. G., Sousa, V. F. C. and Ferreira, B. O. (2023). The impact of the deterioration on wood by chlorine: An experimental study, Materials, 16: 969. doi: 10.3390/ma16030969.

[10] Silva, F. J. G., Fernandes, A. J. S., Costa, F. M., Teixeira, V., Baptista, A. P. M. and Pereira, E. (2003). Tribological behaviour of CVD diamond films on steel substrates. Wear, 255: 846–853. doi: 10.1016/S0043-1648(03)00145-5.

[11] Fernandes, L., Silva, F. and Alexandre, R. (2019). Study of TiAlN PVD coating on stamping dies used in tinplate food package production. Micromachines, 10: 182. doi: 10.3390/mi10030182.

[12] Sousa, V. F. C. and Silva, F. J. G. (2020). Recent advances in turning processes using coated tools—a comprehensive review. Metals, 10: 170. doi: 10.3390/met10020170.

[13] Sousa, V. F. C., Castanheira, J., Silva, F. J. G., Fecheira, J. S., Pinto, G. and Baptista, A. (2021). Wear behavior of uncoated and coated tools in milling operations of AMPCO (Cu-Be) alloy. Applied Sciences, 11: 7762. doi: 10.3390/app11167762.

[14] Su, J., Yin, L., Qin, L., Ma, N. and Huang, J. (2017). Preparation and performance of ZrAlN anti-reflective coatings for low-emissivity glasses. Ceramics International, 43(17): 14616–14622. doi: 10.1016/j.ceramint.2017.07.105.

[15] Ayadi, A., Bouaouadja, N., Durán, A. and Castro, Y. (2020). Hybrid SiO2–ZrO2 coatings for restoring and repairing glasses damaged by sandblasting. Ceramics International, 46(8), Part A, 10634–10640. doi: 10.1016/j.ceramint.2020.01.067.

[16] Eremin, E. N., Yurov, V. M., Guchenko, S. A., Laurynas, V. Ch. and Kasymov, S. S. (2016). Antifriction superhard coatings for drill bits and boring cutters. Procedia Manufacturing, 152: 608–612. doi: 10.1016/j.proeng.2016.07.663.

[17] Zhao, J., Wang, Y., Han, B., Li, M. and Cui, G. (2017). Antifriction effects of Cu2S film on Ni-based MMC coating. Surface and Coatings Technology, 315: 391–398. doi: 10.1016/j.surfcoat.2017.03.001.

[18] Li, J., Gan, L., Liu, Y., Mateti, S., Lei, W., Chen, Y. and Yang, J. (2018). Boron nitride nanosheets reinforced waterborne polyurethane coatings for improving corrosion resistance and antifriction properties. European Polymer Journal, 104: 57–63. doi: 10.1016/j.eurpolymj.2018.04.042.

[19] Lixia, Y., Ying, L., Guannan, L., Zhenghui, L. and Guixiang, W. (2015). Rare Metal Materials and Engineering, 44(1): 28–31. doi: 10.1016/S1875-5372(15)30006-0.

[20] Sousa, V. F. C., Silva, F. J. G., Alexandre, R., Fecheira, J. S. and Silva, F. P. N. (2021). Study of the wear behaviour of TiAlSiN and TiAlN PVD coated tools on milling operations of pre-hardened tool steel. Wear, 476: 203695. doi: 10.1016/j.wear.2021.203695.

[21] Fernandes, L., Silva, F. J. G., Andrade, M. F., Alexandre, R., Baptista, A. P. M. and Rodrigues, C. (2017). Improving the punch and die wear behavior in tin coated steel stamping process. Surface & Coatings Technology, 332: 174–189. doi: 10.1016/j.surfcoat.2017.06.086.

[22] Silva, F., Martinho, R., Andrade, M., Baptista, A. and Alexandre, R. (2017). Improving the wear resistance of moulds for the injection of glass fibre–reinforced plastics using PVD coatings: A comparative study. Coatings, 7: 28. doi:10.3390/coatings7020028.

[23] Uglov, V. V., Khodasevich, V. V., Rusalsky, D. P., Koeniger, A., Hammerl, K. and Rauschenbach, B. (1997). Combination of ion implantation and film deposition for forming an antifriction wear-resistant carbon coating, Surface and Coatings Technology, 93(2-3): 331–334. doi: 10.1016/S0257-8972(97)00070-4.

[24] Barker, A. J., Douglas, T. A., Alberts, E. M., Fernando, P. U. A. I., George, G. W., Maakestad, J. B., Moores, L. C. and Saari, S. P. (2022). Influence of chemical coatings on solar panel performance and snow accumulation. Cold Regions Science and Technology, 201: 103598. doi: 10.1016/j.coldregions.2022.103598.

[25] Adak, D., Bhattacharyya, R. and Barshilia, H. C. (2022). A state-of-the-art review on the multifunctional self-cleaning nanostructured coatings for PV panels, CSP mirrors and related solar devices. Renewable and Sustainable Energy Reviews, 159: 112145. doi: 10.1016/j.rser.2022.112145.

[26] Ko, K., Yoon, D., Yang, S. C. and Lee, H. S. (2021). Brush-painted superhydrophobic silica coating layers for self-cleaning solar panels, Journal of Industrial and Engineering Chemistry, 106: 460–468. doi: 10.1016/j.jiec.2021.11.023.

[27] Rudzikas, M., Šetkus, A., Stange, M., Ulbikas, J. and Ulyashin, A. (2020). Simple interference based colorization of Si based solar cells and panels with ITO/SiNx:H double layer antireflective coatings. Solar Energy, 207: 218–227. doi: 10.1016/j.solener.2020.06.091.

[28] Alim, M. A., Tao, Z., Saeed, N., Hao, X., Abden, M. J. and Rahman, A. (2022). Effect of reflective coating on thermal and electrical performances of solar roof tiles. Energy Conversion and Management, 270: 116251. doi: 10.1016/j.enconman.2022.116251.
[29] Silva, F. J. G. (2017). Nanoindentation on tribological coatings. pp. 111–133. *In*: Atul Tiwari and Sridhar Natarajan (Eds.). Applied Nanoindentation in Advanced Materials. John Wiley & Sons Ltd., Hoboken, NJ, USA, 2017. ISBN: 9781119084495.
[30] Wang, Z., Wang, Z., Zhang, T., Guo, W., Dai, H. and Ding, K. (2022). Reliability evaluation of thermal barrier coatings for engine combustion chambers based on Monte-Carlo simulation. Surface and Coatings Technology, 448: 128923. doi: 10.1016/j.surfcoat.2022.128923.
[31] Huang, Y., He, Y., Yang, B., Khan, A., Zhao, X. and Song, P. (2023). Effect of YSZ/TaSi2-MCrAlY thermal barrier coatings on oxidation resistance in air and water vapor. Ceramics International, 49(9, Part A): 14551–14562 doi: 10.1016/j.ceramint.2023.01.045.
[32] Xiang, Y., Yan, K., Yu, H., Guo, Y., Ying, Y., Li, Z., Sun, J. and Fang, C. (2023). Comparative investigation on the hot corrosion failure of YSZ and GdYb-YSZ double-ceramic-layer thermal barrier coatings under Na2SO4+V2O5 molten salts. Ceramics International, 49(11): 18678–18688. doi: 10.1016/j.ceramint.2023.02.245.
[33] Cen, L., Qin, W. Y., Yu, Q. M. and Chen, F. R. (2022). Thermal barrier coating crack driving force investigation on the basis of a new constitutive framework. Engineering Fracture Mechanics, 275: 108832. doi: 10.1016/j.engfracmech.2022.108832.
[34] Dai, J., Huang, B., He, L., Mu, R., Tian, H. and Xu, Z. (2023). Thermal cycling behavior and failure mechanism of Yb2O3-doped yttria-stabilized zirconia thermal barrier coatings. Materials Today Communications, 34: 105409. doi: 10.1016/j.mtcomm.2023.105409.
[35] Blanchard, F., Kadi, M. J., Bousser, E., Baloukas, B., Azzi, M., Klemberg-Sapieha, J. E., et al. (2023). Effect of thermal ageing on the optical properties and pore structure of thermal barrier coatings. Surface and Coatings Technology, 452: 129080. doi: 10.1016/j.surfcoat.2022.129080.
[36] Arenas, M. A., Conde, A., García, I. and de Damborenea, J. J. (2022). PVD hard coatings on ceramic tiles for aesthetic applications: surface characterisation and corrosion properties. Ceramics International, 48(15): 21794–21802. doi: 10.1016/j.ceramint.2022.04.163.
[37] Holmberg, K. and Matthews, A. (1994). Coatings Tribology—Properties, Techniques and Applications in Surface Engineering (1st Edition), Elsevier, Amsterdam, The Netherlands. ISBN: 9780080875927.
[38] Constantin, R. and Miremad, B. (1999). Performance of hard coatings, made by balanced and unbalanced magnetron sputtering, for decorative applications. Surface and Coatings Technology, 120-121, 728–733. doi: 10.1016/S0257-8972(99)00366-7.
[39] He, J., Lan, X., Liu, Z., Jiao, D., Zhong, X., Cheng, Y., et al. (2021). Modification of Cr/CrN composite structure by Fe addition and its effect on decorative performance and corrosion resistance, Ceramics International, 47(17): 23888–23894. doi: 10.1016/j.ceramint.2021.05.097.
[40] Gupta, P., Fang, F., Rubanov, S., Loho, T., Koo, A., Swift, N., et al. (2019). Decorative black coatings on titanium surfaces based on hard bi-layered carbon coatings synthesized by carbon implantation, Surface and Coatings Technology, 358: 386–393. doi: 10.1016/j.surfcoat.2018.11.060.
[41] Shimpi, K. C., Ravindranath, K., Jani, A. K., Kothari, D. C. and Harindranath, C. S. (1997). Decorative coatings produced using combination of reactive arc evaporation and magnetron sputtering. Surface and Coatings Technology, 90(1-2): 115–122. doi: 10.1016/S0257-8972(96)03102-7.
[42] Ferreira, R., Carvalho, Ó., Sobral, L., Carvalho, S. and Silva, F. (2020). Influence of morphology and microstructure on the tribological behavior of arc deposited CrN coatings for the automotive industry. Surface and Coatings Technology, 397: 126047. doi: 10.1016/j.surfcoat.2020.126047.
[43] Navinšek, B., Panjan, P. and Milošev, I. (1999). PVD coatings as an environmentally clean alternative to electroplating and electroless processes. Surface and Coatings Technology, 116–119: 476–487. doi: 10.1016/S0257-8972(99)00145-0.
[44] Kuhr, M., Bauer, S., Rothhaar, U. and Wolff, D. (2003). Coatings on plastics with the PICVD technology. Thin Solid Films, 442(1-2): 107–116. doi: 10.1016/S0040-6090(03)00956-8.
[45] Zarka, M., Dikici, B., Niinomi, M., Ezirmik, K. V., Nakai, M. and Kaseem, M. (2021). The $Ti_{3.6}Nb_{1.0}Ta_{0.2}Zr_{0.2}$ coating on anodized aluminum by PVD: A potential candidate for short-time biomedical applications. Vacuum 192: 110450. doi: 10.1016/j.vacuum.2021.110450.

[46] Zarka, M., Dikici, B., Niinomi, M., Ezirmik, K. V., Nakai, M. and Yilmazer, H. (2021). A systematic study of β-type Ti-based PVD coatings on magnesium for biomedical applicatio., Vacuum, 183: 109850. doi: 10.1016/j.vacuum.2020.109850.

[47] Staszuk, M., Pakuła, D., Reimann, L., Kloc-Ptaszna, A. and Lukaszkowicz, K. (2023). Structure and properties of the TiN/ZnO coating obtained by the hybrid method combining PVD and ALD technologies on austenitic Cr-Ni-Mo steel substrate. Surfaces and Interfaces, 37: 102693. doi: 10.1016/j.surfin.2023.102693.

[48] Hassan, S., Nadeem, A. Y., Ali, M., Ali, M. N., Niazi, M. B. K. and Mahmood, A. (2022). Graphite coatings for biomedical implants: A focus on anti-thrombosis and corrosion resistance properties. Materials Chemistry and Physics, 290: 126562. doi: 10.1016/j.matchemphys.2022.126562.

[49] Sousa, V. F. C. and Silva, F. J. G. (2020). Recent advances on coated milling tool technology—a comprehensive review. Coatings, 10(3): 235. doi: 10.3390/coatings10030235.

[50] Ibrahim, R. N., Rahmat, M. A., Oskouei, R. H. and Raman, R. K. S. (2015). Monolayer TiAlN and multilayer TiAlN/CrN PVD coatings as surface modifiers to mitigate fretting fatigue of AISI P20 steel. Engineering Fracture Mechanics, 137: 64–78. doi: 10/1016/j.engfracmech.2015.01.009.

[51] Casais, R., Baptista, A. M., Silva, F. J., Andrade, F., Sousa, V. and Marques, M. J. (2022). Experimental study on the wear behavior of B_4C and TiB_2 monolayered PVD coatings under high contact loads. The International Journal of Advanced Manufacturing Technology, 120: 6585–6604. doi: 10.1007/s00170-022-09182-4.

[52] Shen, Z., He, L., Xu, Z., Mu, R. and Huang, G. (2019). LZC/YSZ double layer coatings: EB-PVD, microstructure and thermal cycling life. Surface and Coatings Technology, 367: 86–90. doi: 10.1016/j.surfcoat.2019.03.079.

[53] Silva, F. J. G., Baptista, A. P. M., Pereira, E., Teixeira, V., Fan, Q. H., Fernandes, A. J. S. and Costa, F. M. (2002). Microwave plasma chemical vapour deposition diamond nucleation on ferrous substrates with Ti and Cr interlayers. Diamond and Related Materials, 11: 1617–1622. doi: 10.1016/S0925-9635(02)00029-8.

[54] Mendibide, C., Steyer, P., Fontaine, J. and Goudeau, P. (2006). Improvement of the tribological behaviour of PVD nanostratified TiN/CrN coatings—An explanation. Surface and Coatings Technology, 201(7): 4119–4124. doi: 10.1016/j.surfcoat.2006.08.013.

[55] Martinho, R. P., Silva, F. J. G., Alexandre, R. J. D. and Baptista, A. P. M. (2011). TiB_2 Nanostructured Coating for GFRP Injection Moulds. Journal of Nanoscience and Nanotechnology, 11: 5374–5382. doi: 10.1166/jnn.2011.3772.

[56] Baker, M. A., Rebholz, C., Leyland, A. and Matthews, A. (2002). Electron spectroscopic studies of nanocomposite PVD TiAlBN coatings. Vacuum, 67(3-4): 471–476. doi: 10.1016/S0042-207X(02)00233-6.

[57] Xu, Z. H., Zhou, X., Wang, K., Dai, J. W. and He, L. M. (2014). Thermal barrier coatings of new rare-earth composite oxide by EB-PVD. Journal of Alloys and Compounds, 587, 126–132. doi: 10.1016/j.jallcom.2013.10.157.

[58] Silva, F. J. G., Fernandes, A. J. S., Costa, F. M., Baptista, A. P. M. and Pereira, E. (2005). Unstressed PACVD diamond films on steel pre-coated with a composite multilayer. Surface & Coatings Technology, 191: 102– 107. doi: 10.1016/j.surfcoat.2004.02.036.

[59] Silva, F. J. G., Fernandes, A. J. S., Costa, F. M., Baptista, A. P. M. and Pereira, E. (2004). A new interlayer approach for CVD diamond coating of steel substrates. Diamond and Related Materials, 13: 828–833. doi: 10.1016/j.diamond.2003.10.081.

[60] Tu, R., Jiang, M., Yang, M., Ji, B., Gao, T., Zhang, S. and Zhang, L. (2023). Effects of gradient structure and modulation period on mechanical performance and thermal stress of TiN/TiSiN multilayer hard coatings. Materials Science and Engineering: A, 866: 144696. doi: 10.1016/j.msea.2023.144696.

[61] Zhu, P., Li, P., Ge, F. and Huang, F. (2023). Effect of residual stress on the wear behavior of magnetron sputtered V–Al–N coatings deposited at the substrate temperature < 200°C. Materials Chemistry and Physics, 296: 127218. doi: 10.1016/j.matchemphys.2022.127218.

[62] Ng, C.-H., Rao, J. and Nicholls, J. (2020). The role of PVD sputtered PTFE and Al2O3 thin films in the development of damage tolerant coating systems. Journal of Materials Research and Technology, 9(1): 675–686. doi: 10.1016/j.jmrt.2019.11.009.

[63] Raaif, M. (2019). Investigating the structure and tribo-mechanical performance of PVD TiN on bearing TiN substrate constructed by rf plasma. Materials Chemistry and Physics, 224: 117–123. doi: 10.1016/j.matchemphys.2018.12.004.

[64] Shumilin, A. I. and Fomin, A. A. (2023). Tantalum-containing coatings with superhard inclusions produced by DC-magnetron sputtering. International Journal of Refractory Metals and Hard Materials, 112: 106133. doi: 10.1016/j.ijrmhm.2023.106133.

[65] Pusch, C., Hoche, H., Berger, C., Riedel, R., Ionescu, E. and Klein, A. (2011). Influence of the PVD sputtering method on structural characteristics of SiCN-coatings—Comparison of RF, DC and HiPIMS sputtering and target configurations. Surface and Coatings Technology, 205: S119–S123. doi: 10.1016/j.surfcoat.2011.04.095.

[66] Sousa, V. F. C., Silva, F. J. G., Lopes, H., Casais, R. B. C., Baptista, A., Pinto, G., et al. (2021). Behavior and machining performance of TiAlSiN-Coated tools obtained by dc MS and HiPIMS: A comparative study. Materials, 14: 5122. doi: 10.3390/ma14185122.

[67] Grigoriev, S., Vereschaka, A., Zelenkov, V., Sitnikov, N., Bublikov, J., Milovich, F., et al. (2022). Specific features of the structure and properties of arc-PVD coatings depending on the spatial arrangement of the sample in the chamber. Vacuum, 200: 111047. doi: 10.1016/j.vacuum.2022.111047.

[68] Biava, G., Siqueira, I. B. A. F., Vaz, R. F., de Souza, G. B., Jambo, H. C. M., Szogyenyi, A., et al. (2022). Evaluation of high temperature corrosion resistance of CrN, AlCrN, and TiAlN arc evaporation PVD coatings deposited on Waspaloy. Surface and Coatings Technology, 438: 128398. doi: 10.1016/j.surfcoat.2022.128398.

[69] Wang, B., Jiang, C. and Guo, H. (2022). Effect of sample rotation speed on microstructure and thermal cycling behaviour of (Gd0.8Yb0.2)2Zr2O7/YSZ dual-layer thermal barrier coatings prepared by EB-PVD. Ceramics International, 49(9): Part A, 13559–13569. doi: 10.1016/j.ceramint.2022.12.232.

[70] Grigoriev, S., Vereschaka, A., Zelenkov, V., Sitnikov, N., Bublikov, J., Milovich, F., et al. (2022). Investigation of the influence of the features of the deposition process on the structural features of microparticles in PVD coatings. Vacuum, 202: 111144. doi: 10.1016/j.vacuum.2022.111144.

[71] Martinho, R. P., Silva, F. J. G. and Baptista, A. P. M. (2007). Wear behaviour of uncoated and diamond coated Si_3N_4 tools under severe turning conditions. Wear, 263: 1417–1422. doi: 10.1016/j.wear.2007.01.048.

[72] Paramanik, D. D. B. (2022). Synthesis of nanocrystalline diamond embedded diamond-like carbon films on untreated glass substrates at low temperature using ($C_2H_2 + H_2$) gas composition in microwave plasma CVD. Applied Surface Science, 579: 152132. doi: 10.1016/j.apsusc.2021.152132.

[73] Tau, O., Lovergine, N. and Prete, P. (2023). Adsorption and decomposition steps on Cu(111) of liquid aromatic hydrocarbon precursors for low-temperature CVD of graphene: A DFT study. Carbon, 206: 142–149. doi: 10.1016/j.carbon.2023.02.011.

[74] Govindarasu, K. G., Venkatesan, R., Eswaran, M. and Arumugam, P. (2022). Simple and efficient CVD synthesis of graphitic P-doped 3D cubic ordered mesoporous carbon at low temperature with excellent supercapacitor performance. Advanced Powder Technology, 33(3): 103439. doi: 10.1016/j.apt.2022.103439.

[75] Boyle, H., Norgren, S., Crawforth, P., Christofidou, K., Boing, D. and Jackson, M. (2022). Insights in α-Al2O3 degradation in multilayer CVD coated carbide tools when turning IN718. Wear, 523: 204786. doi: 10.1016/j.wear.2023.204786.

[76] Lofaj, F., Kabátová, M., Kvetková, L., Dobrovodský, J. and Girman, V. (2019). Hybrid PVD-PECVD W-C:H coatings prepared by different sputtering techniques: The comparison of deposition processes, composition and properties, Surface and Coatings Technology, 375: 839–853. doi: 10.1016/j.surfcoat.2019.07.078.

[77] Martinho, R. P., Silva, F. J. G., Martins, C. and Lopes, H. (2019). Comparative study of PVD and CVD cutting tools performance in milling of duplex stainless steel. The International Journal of Advanced Manufacturing Technology, 102: 2423–2439. doi: 10.1007/s00170-019-03351-8.

[78] Silva, F. J. G., Martinho, R. P., Martins, C., Lopes, H. and Gouveia, R. M. (2019). Machining GX2CrNiMoN26-7-4 DSS Alloy: Wear Analysis of TiAlN and TiCN/Al2O3/TiN Coated Carbide Tools Behavior in Rough End Milling Operations. Coatings, 9: 392. doi:10.3390/coatings9060392.

CHAPTER 8

Greener Technologies in Other Manufacturing Processes

8.1 The importance of being greener in any type of industry

In view of the unavoidable climate change and severe weather phenomena that one has been witnessing more and more frequently in recent years, it is extremely urgent to act assertively and effectively, with a view of trying to reverse the chaotic situation that is foreseen for the coming decades. Environmental awareness is still quite heterogeneous in world terms, largely due to the education and awareness of populations to the causes that are at the origin of the problems, and that are beginning to affect their lives in an increasingly intense and violent way. The scientific community in general, and environmental organizations in particular, have made an additional effort to generate the necessary alerts to change some usual behaviors. If every consumer actively contributes to the problem, it is also up to them to contribute to the solution, trying to consume less and be more selective in their consumption options, giving priority to products manufactured according to procedures that respect the environment as much as possible. Below are some more conceptual approaches regarding recent efforts to make manufacturing operations more environmentally friendly.

8.2 Conceptual developments about greener technologies in industry

The use of greener processes is essentially based on technological innovation and the availability of companies to adopt these new technologies that are less harmful to the environment. Both tasks are arduous. Technological innovation has a price, which should be passed on to products to make companies economically viable. This establishes new rules of competitiveness in the market: if, on the one hand, the product provides greater advantages in environmental terms, on the other hand, it usually presents a higher cost, which can direct these products to a class of consumers more receptive to the essentially related need to preserve the environment. However, the market is perfectly aware that the number of consumers who make their consumption choices concerned with the environment is still quite limited in most societies, which largely restricts the flow of products manufactured with this type of concern and which have a higher price for that reason.

In a study carried out having as the target, Pakistan, Shahzad et al. [1] concluded that the trend towards the adoption of green innovation is greater, when the perception is higher and intention of managers comes to adhere to this type of technological innovation focused on the environment. These authors also stated that the reality is quite different considering the size of the organizations. It was also clear in the study that facilitating access to green innovation by governments and public bodies, through incentive programs, is a determining factor in increasing the number of companies and organizations that plan the adoption of green innovation in the future.

Song et al. [2] published a much broader study, relating various aspects connected to the adoption of best environmental practices by companies, namely: environmental regulation, staff quality, efficiency of the R&D team, green technology, and profitability, based on the consultation of 1197 Chinese companies in 16 different industrial sectors. The results obtained through mathematical models are not completely conclusive, as each model has presented analysis limitations due to the large number of factors that can influence this type of study. However, through this study it was possible to identify some concerns about homogenizing the sample in terms of the location of the companies, the type of market in which they normally operate, among other factors. Some important ideas emerge from the study: (a) the increase in government support for the adoption of more environmentally friendly policies will encourage companies to invest more quickly in systems that allow the reduction of their greenhouse gas emissions; (b) hiring and training its staff to a higher level will contribute to much more efficient energy consumption; (c) the adoption of green technologies will increase companies' commercial yields, although it is not yet known exactly what level of adoption of green technologies will allow companies to maximize their profits. This study was carried out in China, and was essentially based on industrial companies, which left room for new studies to be carried out involving companies that essentially provide services to industrial companies, and on which the industry also increasingly depends.

Gao et al. [3] addressed the problem of adopting green technologies in industry, but from a different and broader perspective. In fact, there are numerous external factors that negatively affect the evolution of companies towards a broader adoption of green technologies, inducing a lack of motivation for this change. However, their behavior in choosing the path to adopt is also strongly influenced by other external factors, which will be discussed below. These authors divided green technological innovation into three aspects, depending on the behavior of companies regarding its adoption. Class I corresponds to a passive adoption, essentially induced by external factors, not convinced, and which essentially aims at saving resources. In Class II, the company's involvement becomes active, i.e., not only dependent on external factors, but also induced by its own convictions linked to environmental preservation, not having as its main objective only the saving of resources, but also trying to contribute in a way proactively towards environmental improvement. Class III is defined as a mix of passive and active adoption, in which the company tries in every way to be more environmentally friendly, due to strong convictions in that sense, and not with the objective of saving resources. Obviously, the position of companies also depends a lot on the competitive context in which they are inserted, the added value they usually generate, the marketing associated with them, the mentality of the top management, the well-defined mission of the company, and the culture that is fostered within the company regarding the environment. The communication factor is also an extremely important vehicle for making the main guidelines established by top management flow down to the most basic operational ones. On the other hand, green technological innovation can be encouraged along three different lines: command-control, market-based incentives, and public participation. Command-control derives essentially from regulation and government requirements for the fulfillment of certain national or regional objectives, in which the regulation establishes certain rules (command) and establishes mechanisms to monitor their application by companies (control). Market-based initiatives are essentially conditioned by the performance of competitors, in which competitors' marketing can induce a response from other companies in order not to lose market share. Indeed, advertisements relating to greener products or production policies could induce part of the market to opt for these products, which requires an adequate response from competitors, with a view of accompanying these initiatives and not losing potential customers who feel attracted by products and campaigns of those who have adopted a more environmentally friendly way. Finally, public participation takes place in more developed societies and represents a mass motivation for more environmentally friendly products, requiring companies to respond to their desire to preserve the environment.

After defining the different factors to be considered in the analysis, the authors carried out a very interesting study that allowed establishing several very assertive conclusions and scientific relevance. This study concluded that the adoption of green manufacturing technologies normally generates an increase in revenue from sales (economic aspect), improves worker satisfaction through an increase in awareness that they are producing something that better preserves nature (social aspect), and globally promotes a reduction in environmental degradation caused by industrial activity (environmental aspect). On the other hand, among the three aspects identified as an incentive for the adoption of green technological innovation, the command-control aspect, when applied abruptly and massively, proved to be harmful to the adoption of green technological innovation, since it requires large investments and significantly degrades the commercial yields of companies, which is seen as a harmful action for development by companies. Thus, this action should be applied moderately in an initial phase, being then more incisive as time goes on, in order to encourage in a first phase, and to condition the operation in a second phase, putting competitors on an equal footing in fulfilling environmental duties, but without causing abrupt initial investment needs, which could cause economic turbulence and lack of perception of objectives by companies. Furthermore, a more moderate application of the obligations to which companies may be subjected by government regulation can bring added economic benefits to companies, which can also translate into a shift from Class I to Class II regarding the adoption of green technologies innovation, i.e., stop adopting green innovation in a passive way, to adopt it with conviction. Moreover, it is expected that the losses registered by a more tenuous command-controlled environmental performance can be compensated by the adjustment of other environmental regulations. The study also led to the conclusion that high-intensity market-driven environmental regulations can significantly promote the transformation of green technological innovation in manufacturing industries, a factor that will tend to promote a continuous transformation of the industry in accordance with the principles of green technological innovation. On the other hand, the study also concluded that voluntary environmental regulations do not significantly impact the adoption of green manufacturing, having practically no influence on green technological innovation. This conclusion, combined with the earlier one, points to the greater effectiveness of moderate but mandatory regulations. Only in this way can effective results be achieved. Given that these technological changes are complex and usually involve different aspects of engineering, it has also been proven that acting on account of only one scientific area of engineering can be a blunder. Effectively, the intersection of knowledge in the various areas of engineering produces a much more effective combined effect, as there is a complementarity of solutions in the various areas to overcome problems that are often multidisciplinary. The use of different tools helps a lot to overcome the difficulties that most of these processes present.

Peng et al. [4], in a study carried out based in the Chinese province of Jiangsu, drew some noteworthy conclusions regarding how governments and companies should face the adoption of green technological innovation. The first conclusion is related to the number and quality of patents related to green innovation, which should be largely encouraged. Effectively, governments should encourage programs that attract the best students, creating an environment for reinventing technological processes that essentially aim to increase energy efficiency, reduce harmful effects on the environment and reduce the overall environmental impact, including the reduction of sewage. This second factor on which the study focuses, clearly mentioning that the government applies excessively moderate fees to the companies that generate the highest volumes of sewage. Effectively, if these companies were charged with larger fees, their profits would be harmed, which would encourage companies to adopt a new sewage generation policy, investing in the reduction of sewage production to increase their profits, which would be extremely beneficial for the environment. The study also calls for a change in the recruitment system, encouraging the recruitment of students who identify with green technological innovation policies, attracting the best talent and encouraging training in this area at the universities to be strengthened. On the other hand, companies should take advantage of government financial incentives that are distributed to projects in the technological field, to encourage the development of knowledge and innovation in green technology. If these principles were followed by companies and governments, the level of green technological innovation would certainly be much higher, and manufacturing processes would be much more developed and environmentally friendly.

Among other conclusions like to those highlighted earlier in other works, and also having China as the basis of their study, Li et al. [5] emphasized that government policy should discourage manufacturing processes that are more polluting, in the form of carbon emissions, being much more restrictive with this type of industries and helping industries that emit less carbon, which should also be responsible for their own incentive programs to reduce carbon emissions. In the case of the most polluting companies, the government should force companies to abandon carbon-emitting processes, through technological innovation and the adoption of new manufacturing processes for the same type of products, or even abandon the manufacture of products that largely use carbon-emitting processes.

Zhu et al. [6] studied the causes that could be behind the adoption of different modes of green manufacturing, both in terms of manufacturing, re-manufacturing, and the hybrid mode, also having the Chinese market as the scenario. The conclusion is in line with what was mentioned above: the main external influence is taxes on the pollution generated and emission quotas stipulated by the government, while as internal causes the gaps in terms of R&D are pointed out in relation to the costs of the technology involved in the re-manufacture. Regarding fees and quotas established by governments, it seems almost consensual that they are low enough to allow companies to continue polluting, as the costs of innovating and changing current technology for more evolved green technology seem to be higher than what is required of them in terms of fees. Meanwhile, companies that do not have top managers sensitive to environmental issues will continue to use technology that is not friendly enough to the environment, because it is economically favorable to them, and because it is more familiar to their workers. It should be remembered that new technologies imply training in the operation of new equipment, updating knowledge and maintenance capacity for these new technologies, situations that bring inconvenience to the normal flow of production during a transition period, although it is foreseeable that this will have to happen in the very near future, by the imposition of regulations. As for the technology and advantages of re-manufacturing, a deeper investigation is still needed to balance and optimize this type of operations in economic terms, given that there are significant gains in environmental terms. However, the uncertainty in the state of reception of products for re-manufactures, still raises many doubts regarding the process and economic profitability in this way. Hence, re-manufacturing is still a subject under intense study, being still difficult to parameterize, given the uncertainties mentioned above.

Liu et al. [7] compared the difficulties that affect the adoption of green manufacturing in China, Germany, and the USA, stating that there are differences. A common problem is the gap in establishing products that can be taken as baselines for comparison and evolution in the process of greening production. As for the differences, the authors state that the close vigilance on the part of the Chinese authorities regarding the evolution of the implementation of green technologies has produced significant results, which is gradually helping the industry in China to adopt the principles needed to make production systems more environmentally friendly. As for Germany and the USA, it becomes more difficult to carry out detailed analyses, given that production is often distributed, i.e., products are not produced from start to finish within the same company or plant, which leads to difficulties in identifying which company is responsible for each process. The authors also noted that analyzes of the life cycle of products are rarely carried out, including their use phase. This gap also leads to serious difficulties in establishing comparisons and in producing results that are sufficiently reliable to allow a more accurate perception of the strategies being implemented, as well as the impact of these strategies on the environment. These are limitations relatively common in the different countries analyzed, in addition to the limitations presented by many companies in accurately quantifying the environmental impact of their manufacturing processes, mainly when several companies are involved and with different duties, considering that outsourcing is quite common in Western countries. In fact, an analysis of the product overall could involve analyzing the processes of several companies.

Serrano-García et al. [8] also carried out a study to analyze the adoption of green technological innovation, but with Europe as scenario. To this end, 1018 companies from seven European countries were contacted. The study concluded that: (a) the adoption of Green Production Capabilities and the Technical Dimension, when implemented individually, even if intensively, have a significant impact in terms of environmental and financial performance. However, when they are implemented in association, their contribution to environmental performance is evident, but the same cannot be said for financial performance. In addition, when this association is implemented on a massive scale, it turns out that there is no significant effect either on the environment or on financial performance. Therefore, depending on the focus of the application of each of the aspects considered in the study, it becomes necessary to take due care to implement only one of the aspects, or in a moderate way both aspects, since the joint application and in massive doses, do not present any advantage. These results will certainly need to be confirmed by other studies and deepen the reason why these two strands collide in terms of implementation.

After this more general vision of how industry can use innovation to develop processes that produce less environmental impact, it is now time to proceed with some analyzes of four industrial sectors that, due to their importance in our daily lives and economic relevance, deserve to be highlighted in relation to others.

8.3 Making the steel industry greener

The steel industry has enormous relevance in the economic context and is also linked to much of what we consume, namely mobility and construction. In addition, steel is still the material par excellence for building numerous structures and equipment. However, the high melting and processing temperatures of steel require vast amounts of energy, water, and some gasses. Considering the data presented by Patnaik et al. [9] and Ledari et al. [10], the steel industry is the worldwide largest responsible for energy consumption, about 5%, and is also responsible for about 4–7% of CO_2 emissions generated by humanity [9,10]. Moreover, around 80% of the energy consumed worldwide comes from fossil raw materials [11]. This energy sector was also responsible for the emission of about 33.8 billion tons of CO_2 during 2022, which represents an increase of about 2 billion tons compared to the previous year [12]. It should also be mentioned that blast furnaces are the type of furnace most used in the steel industry, and that these furnaces are responsible for around 90% of the entire volume of emissions generated along the steel production chain [10]. In addition to melting, the subsequent rolling operations, with a view of obtaining products such as sheet metal, rods, tubes, etc., also require abysmal amounts of energy, since steel essentially needs to be worked hot. Making an excellent contribution to recycling, it is also an industry that produces severe greenhouse gas emissions. And it is precisely in this field that studies have emerged with a view to reducing these emissions or making them less harmful to the environment.

In a recent study published by Honma et al. [13], it is categorically stated that the replacement of blast furnaces by electric furnaces contributes significantly to the reduction of CO_2 emissions. In addition, a reduction in the coal ratio by replacing it with electricity in the total energy consumption contributes positively to improving efficiency. These authors are quite critical of the carbon taxes applied to industry in Japan. In fact, the carbon tax applied is quite low, being in 2022/2023 of 1.82 €/ton CO_2. Furthermore, coal and coke used in steel production are exempt from taxes. This policy aims to protect the economy of each country, and certainly any country cannot unilaterally raise carbon taxes exaggeratedly, under penalty of making its industry uncompetitive, and having serious market and economic problems. Therefore, a common tax policy must be agreed between the different steel producing countries, avoiding the use of fossil fuels and the emission of CO_2. Otherwise, the inhibition of application of taxes needs to be overcome, and an effective step forward towards to the decarbonization of this industry needs to be finally done. Only a common policy, duly concerted, will be able to create equal conditions in the application of these rates, so as not to create destabilization of the market, but contribute positively to the steel making processes becoming effectively greener. Obviously, the generation of electricity is another problem, which can be overcome by generating energy through renewable means, avoiding the use of fossil products to generate the additional electricity that would be needed under the adoption of electrical furnaces. In that study, the authors carried out an analysis of the volume of CO_2 emissions at a particular interval (2008–2019) and found that it would be possible to reduce these CO_2 emissions by around 60.6% just by applying a series of actions around the steel making industry without negatively effecting the global efficiency of the process. However, the efficiency of replacing coal with electricity decreases with the increasing percentage of the replaced coal. In fact, 10% of coal substituted corresponds to a 23% reduction in CO_2 emissions, but replacing 20% of coal with electricity only corresponds to a 33% reduction in CO_2 emissions, values that will tend to converge. However, the benefits in terms of emissions will be ever-increasing and very welcome. In addition, the authors also studied emissions related to logistical issues. Given that steel has a relatively high specific weight, the transfer of this material between the different stages of production has sufficiently harmful impacts in terms of emissions, need to be reconsidered. Thus, it is possible to conclude that agglomeration in this industrial sector is highly desirable. Moreover, recycling is gaining more and more intensity in the steel industry context. This is beneficial in two ways, as it avoids the landfill of products and generates less emissions than using ore for steel production. However, the collection scrap centers must be close to melting, and this should be close to the rolling processes, avoiding transfer and the corresponding emissions between these manufacturing sectors.

However, technology is changing, and it is possible to use Green Hydrogen as an energy source for steel production. Effectively, hydrogen is considered the most promising energy source of the 21st century, given that it is a clean source that can have a number of sources from which it comes, has a high calorific value, high thermal conductivity, and high reactivity [10]. Effectively, like carbon monoxide, hydrogen can reduce iron oxides, giving rise only to water, to the detriment of the production of CO_2. Incidentally, hydrogen is considered the most efficient reactor among those that are known and can be used for industrial purposes [14]. According to predictions made by Elmquist et al. [15] in 2002, the potential for reducing CO_2 emissions in the steel industry was around 2.3 GTon/year, considering the worldwide activity of this sector. Nowadays, some technologies for the use of green hydrogen in the steel industry are already well defined, namely Hydrogen Plasma Reduction (HPSR), Hydrogen Novel Flash Ironmaking Technology (FIT), Hydrogen Blast Furnace Iron making (HBFI), and Axial Furnace Direct Hydrogen Reduction (H-DR) technologies [16]. The first two technologies consume only hydrogen, but the technology is still in a maturation phase, as it has only been verified on a laboratory scale. The technology of Direct Reduction (H-DR) of iron oxides using hydrogen is considered the most promising technology, and could be the source of a 91% reduction in CO_2 emissions when natural gas was completely replaced by hydrogen in this function [17]. Proof that the use of hydrogen has assumed significant importance for the steel industry and for the governments of the most industrialized countries, and that it is seen as a viable and highly desirable alternative on the way to decarbonizing this sector as much as possible, several programs have been launched by different countries with a view to finding effective solutions that can make this industry greener. These programs have assumed different names, such as the AISI program (USA), COURSE (Japan), Carbon2Chem (Germany), HYBRIT (Sweden) and ULCOS (Europe) [10]. These programs were started between 2015 and 2017 but have not yet reached the industrial maturity required to be used on a large scale, as required in this type of industry. Both through these projects and through agreements between companies and government-businesses, since 2021 there has been a significant volume of initiatives aimed at making this industry greener. One of the limiting factors is the production of green hydrogen in sufficient quantities to feed this industry. With this objective, H2 Green Steel (H2GS) in Sweden, a country with a long tradition in steel production in Europe, has made efforts and found some solutions for the production and use of green hydrogen with a view to its use in steel production, through Direct Reduction Iron (DRI) technology. This is expected to contribute to a 95% reduction in CO_2 emissions, produce around 5 MTon of green steel by 2030, and avoid the emission of 0.3 Bton of CO_2 by 2040. However, the cost of a new installation of this nature cannot be overlooked. In fact, the hydrogen generation installation capable of feeding DRI + EAF (Direct Reduction Iron + Electric Arc Furnace) furnaces costs around 10 B€, for a capacity of 7.5 Mtons/yr. For companies to really be able to decarbonize their processes, there are extremely high costs that will necessarily have to be partially borne by government entities, without distortions of the market in terms of competitiveness. Meanwhile, the company Metso (USA) claims that its Circored technology is the only one to produce steel on an industrial scale using hydrogen, through hydrogen-based direct reduction using fine ore instead of pellets [15]. The US has explored relatively different avenues to make the steel industry greener, namely through two university-linked programs:

Molten Oxide Electrolysis, being developed by The Massachusetts Institute of Technology, and Flash Melting Technology, being developed by the University of Utah. The former seeks to produce iron using molten oxide electrolysis, a method that generates nearly zero GHG emissions. The second one was adapted from mining processes and is essentially based on adapting and improving furnaces with a view to using hydrogen. On the other hand, the company Rio Tinto sold all its coal farms until 2018, believing that hydrogen will be the energy source of the future in the primary phase of steel production. Regarding steel production and given the high CO_2 emissions traditionally generated by Blast Furnaces/Basic Oxygen Furnace, techniques have also emerged that aim to capture the CO_2 produced, storing it with a view to later use. Although it has not yet been developed and implemented on an industrial scale, it could be a way of reusing an effluent that is causing numerous environmental problems and which urgently needs to be reduced as quickly as possible.

In a more realistic analysis, Wang et al. [18] are optimistic about the transition from steel manufacturing to a greener industry, but summarize the needs that this greener industry requires, a fact that should also deserve our attention in the current context: to replace 1% of conventional steel production for green steel, 35 GW of renewable energy, 11 GW of hydrogen electrolyzers, and 1000 square kilometers of land would be required. In addition to these facts, this process is valid essentially for the use of iron ore, an ore not abundant in all countries. Thus, with the growth in demand due to the desired transition, the price of iron ore can rise consistently in the long term, a factor that does not help in the transition. It should also be noted that these methods are not the most suitable for recycling steel, but for this purpose electric furnaces can be used, which can be supplied by renewable energy.

8.4 Making the construction industry greener

Construction is an extremely important sector in the world economy. As construction produces buildings that normally have a much longer lifespan than many other consumer goods, the care placed in this sector should be increased. In this work, some factors related to how construction can be automated have already been mentioned, as well as how to integrate automation into the product itself, the building, with a view to increasing its environmental sustainability throughout its entire life cycle. This chapter is intended essentially to focus on how to make construction processes greener. It is a fact that construction consumes a lot of energy and generates a significant amount of waste, which is often difficult to treat and ends up in landfills. Thus, it will be essentially factors linked to the process that will be addressed here.

Carrying out construction works involves a series of environmental concerns, on which contractors must act as a way to minimize the environmental impact. Effectively, contact with the surroundings can harm the environment in several ways, such as: (a) air pollution; (b) noise pollution; (c) water pollution; (d) waste management; (e) miscellaneous environmental impacts; (f) energy consumption; (g) water consumption; (h) consumption of raw materials.

In a study presented at the beginning of this century, Tam et al. [19] carried out a study based on Hong Kong, in which they included three samples and a very descriptive analysis was made about some precautions that the construction sector should take in its processes. Below, some of those considered as most important are highlighted: (a) minimize casting activities within the building envelope; (b) use internal lifting systems, avoiding the use of external winches; (c) reduce dust emissions through the use of asphalted/concrete pavements into construction sites; (d) extra care with cement storage to prevent it from being carried away by the wind; (e) adopt off-site prefabrication whenever possible; (f) preferably use lightweight aluminum scaffolding; (g) promoting waste recycling; (h) use a wastewater treatment facility; (i) use water sprinkler systems for atmospheric pollutant materials used in spray form; (j) use segregation tanks to construction waste; (k) use movable noise barriers; (l) respect the deadlines stipulated for the execution of the work; (m) use electric poker vibrating system for concreting.

Given that there is specific legislation, it is up to contractors to comply with existing regulations and provide conditions for the management of the work to be carried out in a careful manner. There are numerous indicators that can be permanently analyzed, with a view to correctly monitor the work in environmental terms. Tam et al. [19] developed a tailored Green Construction Assessment tool that allows continuous monitoring of the construction work, through the collection of data and calculation of various indicators that, through pre-established values, allow analyzing the degree of fulfillment of objectives and identifying possible improvement actions, with a view to making the construction process greener.

Environmental care in construction must begin with the project. During this phase, the subsequent construction should be properly thought out and planned, directing contractors to comply with certain requirements initially imposed, and which are the most beneficial for the environment. For this purpose, factors such as the use of green technologies, green procurement, green labeling, green building rate tools, and waste management should be considered. All these factors will need to be transposed into the reality of the project, providing accurate indications of how to follow these principles during the construction process, since each work has its own characteristics, sometimes quite different from others already carried out. The environment itself may condition certain aspects of the construction and the environmental respect that must be observed during the execution of the work. Given that each work is multifaceted and involves a significant number of different companies and specialties, the initial definition of rules and adequate planning of the work is vital for its success. On the other hand, if there are environmental principles to be respected, procurement should be carried out based on the history of compliance with the environmental and technical requirements necessary for the project. The work by Bohari et al. [20], despite being very focused on the reality of Malaysia, touches on the main points to be fulfilled so that the construction industry can follow a greener path.

Darko et al. [21] sought to bring together the main drivers with a view to Green Building Technology adoption. To this end, and having the scenario of the reality in Ghana, they consulted 43 construction professionals and based on 21 driving forces previously identified in literature, they sought to find the top five within these 21 driving forces initially defined. One of the driving forces most mentioned by respondents was "setting a standard for future design and construction". This issue brings up again the subject mentioned above. This can be considered as a specific problem of Ghana and other countries. However, if there is standardization that establishes the guidelines to be followed in the project phase, the following phases of implementation of that same project will have the work much easier, having only to comply with what was defined. Thus, it can be highlighted that the design phase is extremely important, as it also defines the planning of the work, which helps to comply with the environmental requirements that are defined below. If standardization is a fact already rooted in the most developed countries, most developing countries still lack standardization in this sense. This lack of standardization is also felt more in smaller constructions, where there is a natural tendency for the selected contractors not to have the necessary skills in terms of standardization. The creation of standards would facilitate work and communication between those who design, those who plan and those who carry out the work. Another factor most cited by respondents is related to "greater energy efficiency". The management of different work teams, planning the arrival of materials, planning the materials displacement within the construction site, and planning the work itself, are some of the factors that could contribute to inefficiencies in energy consumption throughout the construction process. In this case, careful planning and management of work is the best solution to improve energy efficiency. However, environmental education of all workers involved can largely help in achieving the desired objectives in terms of energy efficiency. Another factor mentioned by respondents is essentially related to safety at work: "improved occupants' health and well-being". In this case, the conditions provided by contractors to their work teams are fundamental. This is another typical problem that does not apply uniformly. If in more developed countries it is unthinkable for any worker to take risks in their work and not have the appropriate conditions for the tasks they are required to perform, in developing countries there is a different culture, in which workers are willing to take risks at work. Regulation is usually much more demanding in developed countries than in developing countries. Risk usually leads to accident. As countries evolve, it becomes increasingly difficult to find workers who are available to take risks of accidents or ill-health at work. Therefore, employers end up being forced to improve working conditions. The adverse conditions that many constructions at present are known for, due to the complexity of the work and the lack of safety systems that would be usual in other conditions. Therefore, here too, it is essential to properly plan the work and, as construction progresses, create immediate safety and comfort conditions for workers, ensuring their well-being. The implications that safety breaches, work accidents or absenteeism from work have on other team members have already been studied at length, so it is essential to create the necessary conditions for the safety and well-being of workers in the naturally adverse conditions of a construction site, mainly that they also respect the safety standards imposed on them, that they do not take unnecessary risks and that they are alert to the possible creation of conditions for accidents, alerting managers to these situations and contributing in a positive way so that accidents are avoided. Another factor mentioned by respondents was "reduced whole lifecycle costs". This issue is essentially related to the project and has already been mentioned in another chapter of this book. Architecture and engineering should have the data, as well as the necessary common sense when choosing materials, taking into account previous studies of the life cycle and respective environmental impact of these materials. The construction owner will certainly have a previously defined budget for the construction work and the target audience the company intends to serve, but there are numerous options within each cost range, which allow choices to be made that are approximately in line with the environmental objectives pursued. The selection of sustainable materials, which provide a longer life without the need for a high number of recovery interventions, the preference for natural products, and care with the impact caused at the end of life, will be the path that should be chosen when there are concerns about making any construction greener. Another highly focused aspect was "non-renewable resources conservation". In this case also, design and planning are fundamental. There is a natural tendency, especially in smaller buildings and subject to less careful planning, for there to be waste in terms of some materials used to support the construction. Proper construction design could allow the use of materials and tools that can be reused several times, with the workers being responsible for taking care of these tools and having the necessary environmental education to preserve the materials. However, careful management of any construction can prevent waste, thus contributing to the reduction of environmental impact. Apart from these five aspects most mentioned in this study, all the others are also important and should be taken into consideration so that there is a true evolution in construction towards greener processes in this sector. Of all the others, "Facilitating a culture of best practice sharing" stands out, which is essentially related to another one, "Attraction and retention of quality employees". Despite construction management being a fundamental pillar for its success, especially in developing countries, workforce training is still relatively scarce, which leads to a natural lack of preparation for many of the situations described above. As in many other sectors, training is essential, made even more essential by the higher level of risk involved and the severity of accidents. Given that in developing countries, construction-related tasks are essentially carried out by people with lower levels of education, training is of vital importance. In developed countries, although these tasks are normally carried out by workers recruited from abroad, there are usually well-established rules that should be followed, as well as adequate training, which prevents a high number of accidents and preserves health and well-being of the workers.

In the area of materials to be used, which is not the focus of this chapter, there have also been extremely interesting efforts in the development of new materials that combine others difficult to recycle, and that do not present toxicity when applied in certain situations. Saeli et al. [22] detailed the use of calcareous sludge in green geopolymers for construction. Calcareous sludge is an alkaline waste provided by the paper pulp industry, which is usually sent to landfills. With this solution, a new constituent is acquired for construction and the disposal of this waste in landfills is avoided. The best compromise between properties and workability, according to the authors, is obtained at 10% of calcareous sludge, improving mechanical resistance by around 30% and not compromising the geopolymer's transformation process. Still based on light geopolymers (bulk density of 1.5–1.6 g/cm^3 and 35% porosity) and highly porous foams (75% porosity) intended for construction, Altamiri et al. [23] added volcanic quarry scraps provided from Italy, which present fine particle size and little market interest, using two distinct alkali activation routes. Despite the characterizations made around these materials, no optimization of the content of volcanic quarry scraps was carried out, nor were other ways of activating them studied, so the study can still be considered embryonic, although promising. Castro et al. [24] also incorporated waste plastics reinforced with fiberglass from pultrusion in polyester resins, obtaining promising results in terms of mechanical characteristics. Depending on the size of the waste, the best results in terms of mechanical strength varied between 4 and 8% of waste in weight. Ribeiro et al. [25] also carried out studies with a view to integrating these same residues into sand aggregates and filler replacements, and the incorporation of silane as a binder was also studied. The study considered the inclusion of 4 to 8% in weight of glass fiber reinforced plastics waste as the limit to not compromise the flexural strength and compressive strength of the modified polyester. Indeed, the best results for the purpose of maximizing both flexural and compressive strengths of modified polyester were achieved for 8% in weight of resin replacement by coarse waste particles and incorporation of 1% in weight of active silane as resin binder. Another similar study [26] using the same base polymer (polyester) and glass fiber reinforced plastics residues was also presented, admitting that up to 12% of these particles that are intended to be recycled could be incorporated, without deterioration of compression and flexural strength values shown by the aggregate. There are many other studies that exist and will be under development in this area, which is of great interest for the environment.

8.5 Making the pulp and paper industry greener

Paper is something that one tries to use less and less, through digitalization, but it is unavoidable that it continues to be present in one's daily lives. This type of industry normally has two harmful aspects for the environment: it contributes to deforestation, as the paper pulp is usually extracted from eucalyptus, and it has adjacent chemical processes that consume a lot of water and produce a lot of effluents that contribute to the degradation of the environment, due to bleaching needs, among other processes. Furthermore, it is also a strong energy consumer. Considering data referring to 2021 published by FAO [27], the paper industry consumes around 35% of roundwood production. Furthermore, it is also responsible for around 6% percent of global energy consumption, and is also a strong emitter of CO_2, being responsible for around 2% of the whole industrial emissions [28,29]. Even considering all the digitalization efforts that are currently being carried out around both companies and the population in general, the growth of the world population in less developed areas should dictate an increase in paper consumption in the coming decades. Given these factors, this is one of the sectors where serious efforts should also be made so as to make it greener. However, this industry has two major advantages: it uses raw materials that are renewable, and the final product is easily recyclable. These factors minimize the negative impacts mentioned above, although they do not cancel them out, obviously. The pulp and paper industry is quite heterogeneous in terms of the degree of development between processes used in developed and developing countries. Hence, the analyzes need to be carried out differently. Furthermore, paper takes different forms, and its manufacturing process differs greatly depending on the desired final product. Paper recycling is also less demanding in terms of energy and is also more environmentally friendly because it generates less effluents and greenhouse gas emissions. However, due to the breakdown of fibers during the recycling process, it is always necessary to add new pulp so that the final product meets the normally intended requirements.

Shang et al. [30] carried out a study on the evolution of the pulp and paper industry between the years 2000 and 2030, concluding that developing countries have evolved more quickly in adopting environmentally friendly production policies than developed countries, although there is still a marked difference in the level of green manufacturing between these groups of countries. Developing countries, clearly led by China, have made significant progress in economic and social terms, but still have gaps in the environmental aspect, due to excessive consumption of energy and resources. A more accurate analysis in the case of China, suggests that it will follow the standards already pursued by the USA, first promoting saturation in the economic aspect, and only then focusing its attention on the ecological aspect. There is still a long way to go for this industry in environmental terms, with developed countries competing to take a leading role in the use of green technologies for producing pulp and paper, continuously and effectively reducing the consumption of energy and other resources, as well as the emission of greenhouse gasses. Developing countries should make an additional effort at convergence, adopting technologies already used by developed countries, following the evolution towards a greener pulp and paper industry. Decarbonizing this industry should be the main focus, which can only be achieved through effective cross-border cooperation.

Luukkanen [31] argued that energy should be effectively labeled, so as to allow the pulp and paper industry to understand its origin. This way, companies could make a more rational option, depending on the companies' objectives and mission, as well as the image they intend to convey to the market. Furthermore, the consumption of renewable energy could allow the products produced (various forms of paper) to also be commercialized with the respective label, indicating to the customer the level of environmental concerns placed by the producer in the production of that type of product. In addition to the end consumer who consumes paper essentially for their home printing, the main consumers of paper are printing houses, publishers, and the packaging industry. Here, the relationship is between companies, but the "green" label must be preserved in the chain of manufacture, transformation, and use of paper, so that the final consumer can have the perception and freedom to choose between environmentally friendly or unfriendly products. Obviously, that work concerns a study carried out in Northern Europe, where consumers' environmental awareness is highly mature, and product labeling can make all the difference in terms of choice by end consumers. In less developed countries, the option is often based essentially on the price, and much less on environmental factors, due to the reduced purchasing power of consumers. Although the work precede to 2003, only after 2020 have some electricity producers began to disclose the origin of the energy that companies, and the general population consume. This decalage illustrates the almost two-decade difference that exists between developed and developing countries. In environmental terms, this difference is a severe lack for a rapid transition to a greener economy. In political terms, a greater approximation between the policies implemented in developed and developing countries is necessary, allowing to minimize sharp asymmetries that currently exist in behavioral terms towards the environment.

Sharma et al. [32] focused their study on water consumption and water pollution resulting from bleaching operations that paper is subjected to in certain types of products. Some of the manufacturing stages release perilous organo-chlorine compounds and noxious organo-chlorine compounds, which are harmful to the environment and human health, as they cause carcinogenic, mutagenic, clastogenic and endocrine disrupting effects. However, it is also possible to use greener manufacturing processes in this case, which do not generate these types of compounds. In a review study presented [32], the authors indicated that there are greener ways to bleach paper without the need for harmful chemicals. Therefore, they pointed to the use of Xylanases as a solution, considered as a hydrolytic "bleach-boosting agents", as well as the versatile oxidative Laccases, which can be used alone or mixed, showing an effective role in paper bleaching. The use of biological systems such as bio-bleaching enzymes, the pulp and paper industries can make their processes greener, paving way for sustainable resources. In fact, tests carried out by Sharma et al. [33] with a mixture of bio-enzymes (xylanase-15U, pectinase-2U, a-amylase-2.5U, protease-2U and lipase- 1.8U), when used in the process with the ideal conditions (temperature of 65°C and pH of 9-9.5 for 90 min, followed by reduced chemical treatment), made it possible to achieve a reduction of around 50% in the use of chemicals in the paper bleaching process, which is quite interesting, as well as improving the mechanical properties of the paper obtained by this method, without compromising the bleaching results. Some authors [32] also presented concerns about deforestation caused by the pulp and paper industry, pointing out the use of agricultural residues, such as millet straw, as an alternative. The quantity of straw produced is very high, and millet straw residues are usually burned in post-harvest operations. This burning is harmful to the environment and human health, so reusing it to make paper would bring a double advantage: reduced consumption of trees and lower pollution risks for the environment and people's health.

Although some of these solutions still present uncompetitive costs, it is necessary to explore these alternative routes for the sake of a cleaner industry and a better future for humanity.

8.6 Making the textile industry greener

The textile industry is permanently present in our daily lives and represents an important pillar of the world economy. Due to the low added value of most clothing, this industry has traveled to different regions around the world, currently focusing on Asian countries where labor costs are lower, or even in European countries, North Africa, and South America. The textile industry comprises three subsectors: spinning, weaving, and clothing. The clothing manufacture receives the raw material ready, limiting itself to giving shape to the fabrics, through what is conventionally called fashion. Environmental problems arise essentially in the spinning and weaving phases, where the choice of raw materials is essential, and in the fabric finishing, bleaching and dyeing phases, where chemical processes abound. Therefore, efforts to make this industry greener are essentially focused on the generation of environmentally friendly yarns, and their coloring, through chemical processes that may have a greater or lesser environmental burden. However, concerns cannot be focused only on manufacturing, as textiles have a relatively short useful life and can pose a problem after their end of life. Therefore, there should also be environmental concerns with textile products after their end of life, preferably allowing their reuse or recycling, and preventing them from necessarily having to go to landfills. Next, some developments in green textile industry are reported and summarized, trying to provide a short overview of the different mainstreams of green development in this industry.

Considering cotton dyeing, Chen et al. [34] studied alternatives to the most common fabric dyeing processes, which invariably went through reactive dyeing, and lead to several environmental problems, namely the fact that the effluents cannot be reused and that their post-treatment is quite complex. In fact, the nature of composition and the amount of dyes contained in the effluents depend on the nature of the fabric, the type of dye used and the extent of fabric-dye binding. Dyes can be classified in to basic, anionic, disperse, mordant, sulfur and vat dyes, depending on their way of application. Basic dyes include cationic dyes and are salts, usually chlorides, in which the dyestuff is the basic or positive radical [35]. Anionic dyes are the direct, acid, and reactive dyes. In Chen's study [34], the main barrier to the reuse of spent reactive dye baths—dye hydrolysis—was eliminated by replacing them by water with non-nucleophilic green solvents as the processing medium. Subsequently, the sorption and fixation of the dye were investigated, with a view to verifying whether they were affected by the composition of the solvent and the new type of bases now used. The results obtained showed that compared to conventional aqueous dyeing, the solvent-based process required up to 40% less dye, 97.5% less base and no inorganic salt. The same baths were used in 10 consecutive cycles, demonstrating that they were reusable, unlike conventional baths. Even under these circumstances, the process proved to be consistent and provide high fidelity and color stability. Based on this number of cycles, a waste reduction of around 99% was estimated, for both organic products and salts. This method, applied to five different monochlorotriazine dyes with different chromophores, resulted in the successful dyeing of cotton. Thus, it was found that the solvent dyeing process has significant potential to not require dye or fabric modifications and does not require pressurization. By implementing the principles of waste reduction, reuse, and recycling of resources, thus paving the way for producing more sustainable textiles.

In another study, Mathew et al. [35] developed a low-cost adsorbent from chitosan and polyacrylic acid, using the layer-by-layer on the coir fiber methodology, through alternating exposure to their aqueous solutions. The coir fiber arranged in layers showed a high load of cationic and anionic dyes in both acidic and alkaline pH. In fact, the load ranged between 70 and 90% for an acidic pH of 3. On the other hand, the released wastewater showed to have 81% less dye content, as well as an 84% reduction in chemical oxygen demand. Thus, it has been proven that the use of natural fibers modified through the layer-by-layer technique can contribute very positively to the treatment of effluents in textile dying operations, thus contributing to a textile industry that is less harmful to the environment.

Working on the materials side, Alay et al. [36] devoted their work to the development of antibacterial knitted fabrics based on a mixture of sustainable and environmentally friendly materials, such as PLA (Polylacticacid), lyocell and chitosan fibers. The fabric sample studied was composed of 80% PLA, 15% lyocell and 5% chitosan, and the manufacturing processes were selected to minimize the amount of greenhouse gas emissions produced. The authors reported that subsequent washing operations during the use phase throughout the life cycle of items produced with this fabric would require less water. With regards to cotton, the new materials would also allow these garments to be washed and ironed at lower temperatures, thus saving energy. Furthermore, as this new fabric does not require pre-treatment, water consumption during the fabric finishing process is also lower, thus also having beneficial impacts on the environment. In addition, the new fabric has antibacterial properties without the need for the use of heavy metal antibacterial agents. Still in relation to cotton, items made with the new material showed a lower rate of degradation, thus prolonging the useful life of garments using these materials. In terms of waste from landfill disposal, compared to the control sample made from cotton, samples of the new fabric showed a lower rate of biodegradation, although these fabrics do not present any toxicity, inducing a minimal effect on germination and biomass formation. Therefore, it can be noted that this new fabric presents much more interesting characteristics in terms of environmental sustainability, thus contributing to a greener textile industry.

The interaction between the chemical industry, materials science, and the chemical industry has seen very interesting developments in the last decade, which have been reflected in immense improvements in the materials and processes used to make the impact of the textile industry less intense on the environment. Among the many studies recently developed in the textile area and focusing on the environment, studies in the area of the development of composite fabrics with thermoelectric, energy collection and sensing properties can be highlighted [37], development of ecological conservation products for improving the flame retardancy, UV protection and mechanical properties of fabrics applied to fabrics from historical pieces [38], and the continuous effort to improve the chemical formulations of products used in dyeing and fabric treatment [39], making them less aggressive to the environment. These are just some of the many efforts that have been and continue to be carried out with a view to minimizing the impact of the textile industry on the environment, with the consequences in terms of climate change that we have all felt often.

References

[1] Shahzada, M., Qu, Y., Rehman, S. U. and Zafar, A. U. (2022). Adoption of green innovation technology to accelerate sustainable development among manufacturing industry. Journal of Innovation & Knowledge, 7: 100231. doi: 10.1016/j.jik.2022.100231.

[2] Song, M., Wang, S. and Sun, J. (2018). Environmental regulations, staff quality, green technology, R&D efficiency, and profit in manufacturing. Technological Forecasting & Social Change, 133: 1–14. doi: 10.1016/j.techfore.2018.04.020.

[3] Gao, J., Feng, Q., Guan, T. and Zhang, W. (2023). Unlocking paths for transforming green technological innovation in manufacturing industries. Journal of Innovation & Knowledge, 8: 100394. doi: 10.1016/j.jik.2023.100394.

[4] Peng, B., Zheng, C., Wei, G. and Elahi, E. (2020). The cultivation mechanism of green technology innovation in manufacturing industry: From the perspective of ecological niche. Journal of Cleaner Production, 252: 119711. doi: 10.1016/j.jclepro.2019.119711.

[5] Li, X., Ruan, T., Hou, K. and Qu, R. (2023). The configuring pathways of green technology advance, organizational strategy and policy environment for realizing low-carbon manufacturing from the perspective of simmelian tie: A qualitative comparative analysis of listed companies in China. Journal of Cleaner Production, 382: 135149. doi: 10.1016/j.jclepro.2022.135149.

[6] Zhu, J., Lu, Y., Song, Z., Shao, X. and Yue, X.-G. (2023). The choice of green manufacturing modes under carbon tax and carbon quota. Journal of Cleaner Production, 384: 135336. doi: 10.1016/j.jclepro.2022.135336.

[7] Liu, Y., Wang, Q., Huang, B., Zhang, X., Wang, X. and Long, Y. (2023). Status and challenges of green manufacturing: Comparative analysis of China and other countries. Resources, Conservation & Recycling, 197: 107051. doi: 10.1016/j.resconrec.2023.107051.

[8] Serrano-García, J., Llach, J., Bikfalvi, A. and Arbeláez-Toro, J. J. (2023). Performance effects of green production capability and technology in manufacturing firms. Journal of Environmental Management, 330: 117099. doi: 10.1016/j.jenvman.2022.117099.

[9] Patnaik, D., Pattanaik, A. K., Bagal, D. K. and Rath, A. (2023). Reducing CO_2 emissions in the iron industry with green hydrogen. International Journal of Hydrogen Energy, 48: 23449–23458. doi: 10.1016/j.ijhydene.2023.03.099.

[10] Ledari, M. B., Khajehpour, H., Akbarnavasi, H. and Edalati, S. (2023). Greening steel industry by hydrogen: Lessons learned for the developing world. International Journal of Hydrogen Energy, In Press. doi: 10.1016/j.ijhydene.2023.06.058.

[11] Nemitallah, M. A., Imteyaz, B., Abdelhafez, A. and Habib, M. A. (2019). Experimental and computational study on stability characteristics of hydrogen-enriched oxy-methane premixed flames. Applied Energy, 250: 433–443. doi: 10.1016/j.apenergy.2019.05.087.

[12] Tang, J., Chu, M., Li, F., Feng, C., Liu, Z. and Zhou Y. (2020). Development and progress on hydrogen metallurgy. International Journal of Minerals, Metallurgy and Materials, 27: 713–723. doi: 10.1007/s12613-020-2021-4.

[13] Honma, S., Ushifusa, Y., Okamura, S. and Vandercamme, L. (2023). Measuring carbon emissions performance of Japan's metal industry: Energy inputs, agglomeration, and the potential for green recovery reduction. Resources Policy, 82: 103519. doi: 10.1016/j.resourpol.2023.103519.

[14] Dawood, F., Anda, M. and Shafiullah, G. M. (2020). Hydrogen production for energy: an overview. International Journal of Hydrogen Energy, 45: 3847–3869. doi: 10.1016/j.ijhydene.2019.12.059.

[15] Elmquist, S. A., Weber, P. and Eichberger, H. (2002). Operational results of the Circored fine ore direct reduction plant in Trinidad. Stahl Eisen, 122(2): 59–64. doi: N/A.

[16] Tang, J., Chu, M. S., Li, F., Feng, C., Liu, Z. G. and Zhou, Y. S. (2020). Development and progress on hydrogen metallurgy. International Journal of Minerals, Metallurgy and Materials, 27: 713–723. doi: 10.1007/s12613-020-2021-4.

[17] Rechberger, K., Spanlang, A., Conde, A. S., Wolfmeir, H. and Harris, C. (2020). Green hydrogen-based direct reduction for lowcarbon steelmaking. Steel Research International, 91: 1–10. doi: 10.1002/srin.202000110.

[18] Wang, C., Walsh, S. D. C., Weng, Z., Haynes, M. W., Summerfield, D. and Feitz, A. (2023). Green steel: Synergies between the Australian iron ore industry and the production of green hydrogen. International Journal of Hydrogen Energy, 48(83): 32277–32293. doi: 10.1016/j.ijhydene.2023.05.041.

[19] Tam, C. M., Tam, V. W. Y. and Tsui, W. S. (2004). Green construction assessment for environmental management in the construction industry of Hong Kong. International Journal of Project Management, 22: 563–571. doi: 10.1016/j.ijproman.2004.03.001.

[20] Bohari, A. A. M., Skitmore, M., Xia, B., Teo, M., Zhang, X. and Adham, K. N. (2015). The path towards greening the Malaysian construction industry. Renewable and Sustainable Energy Reviews, 52: 1742–1748. doi: 10.1016/j.rser.2015.07.148.

[21] Darko, A., Chan, A. P. C., Gyamfi, S., Olanipekun, A. O., He and B.-J., Yu, Y. (2017). Driving forces for green building technologies adoption in the construction industry: Ghanaian perspective. Building and Environment, 125: 206–215. doi: 10.1016/j.buildenv.2017.08.053.

[22] Saeli, M., Senff, L., Tobaldi, D. M., Carvalheiras, J., Seabra, M. P. and Labrincha, J. A. (2020). Unexplored alternative use of calcareous sludge from the paper-pulp industry in green geopolymer construction materials. Construction and Building Materials, 246, 118457. doi: 10.1016/j.conbuildmat.2020.118457.

[23] Altimari, F., Lancellotti, I., Leonelli, C., Andreola, F., Elsayed, H., Bernardo, E., et al. (2023). Green materials for construction industry from Italian volcanic quarry scraps. Materials Letters, 333: 133615. doi: 10.1016/j.matlet.2022.133615.

[24] Castro, A. C. M., Ribeiro, M. C. S., Santos, J., Meixedo, J. P., Silva, F. J. G., Fiúza, A., et al. (2013). Sustainable waste recycling solution for the glass fibre reinforced polymer composite materials industry. Construction and Building Materials, 45: 87–94. doi: 10.1016/j.conbuildmat.2013.03.092.

[25] Ribeiro, M. C. S., Meira-Castro, A. C., Silva, F. G., Santos, J., Meixedo, J. P., Fiúza, A., et al. (2015). Re-use assessment of thermoset composite wastes as aggregate andfiller replacement for concrete-polymer composite materials: A casestudy regarding GFRP pultrusion wastes. Resources, Conservation and Recycling, 104: 41–426. doi: 10.1016/j.resconrec.2013.10.001.

[26] Castro, A. C. M., Carvalho, J. P., Ribeiro, M. C. S., Meixedo, J. P., Silva, F. J. G., Fiúza, A., et al. (2014). An integrated recycling approach for GFRP pultrusion wastes: recycling and reuse assessment into new composite materials using Fuzzy Boolean Nets. Journal of Cleaner Production, 66: 420–430. doi: 10.1016/j.jclepro.2013.10.030.

[27] FAO – Food and Agriculture Organization (2021). FAOSTAT database. Retrieved from: https://www.fao.org/faostat/en/#data/FO. (Accessed on July 22nd, 2023).

[28] IEA (Internarional Energy Agency), 2016. International Energy Outlook 2016 with Projection to 2040. IEA Publications, France, 2016. Retrieved from: https://www.eia.gov/outlooks/ieo/pdf/0484. (Accessed on July 23rd, 2023).

[29] IEA, 2021. Tracking industry 2021. Retrieved from: https://www.iea.org/reports/tracking-industry-2021. (Accessed on July 23rd, 2023).

[30] Shang, D., Lu, H., Liu, C., Wang, D. and Diao, G. (2022). Evaluating the green development level of global paper industry from 2000-2030 based on a market-extended LCA model. Journal of Cleaner Production, 380: 135108. doi: 10.1016/j.jclepro.2022.135108.

[31] Luukkanen, J. (2003). Green paper with green electricity? Greening strategies of Nordic pulp and paper industry. Energy Policy, 31: 641–655. PII: S0301-4215(02)00149-0.

[32] Sharma, D., Chaudhary, R., Kaur, J. and Arya, S. K. (2020). Greener approach for pulp and paper industry by Xylanase and Laccase. Biocatalysis and Agricultural Biotechnology, 25: 101604. doi: 10.1016/j.bcab.2020.101604.

[33] Sharma, A., Balda, S., Gupta, N., Capalash, N. and Sharma, P. (2020). Enzyme cocktail: An opportunity for greener agro-pulp biobleachingin paper industry. Journal of Cleaner Production, 271: 122573. doi: 10.1016/j.jclepro.2020.122573.

[34] Chen, L., Wang, B., Ruan, X., Chen, J. and Yang, Y. (2015). Hydrolysis-free and fully recyclable reactive dyeing of cotton in green, non-nucleophilic solvents for a sustainable textile industry. Journal of Cleaner Production, 107: 550–556. doi: 10.1016/j.jclepro.2015.05.144.

[35] Mathew, M. L., Gopalakrishnan, A., Aravindakumar, C. T. and Aravinda, U. K. (2019). Low – cost multilayered green fiber for the treatment of textile industry wastewater. Journal of Hazardous Materials, 365: 297–305. doi: 10.1016/j.jhazmat.2018.11.014.

[36] Alay, E., Duran, K. and Korlu, A. (2016). A sample work on green manufacturing in textile industry. Sustainable Chemistry and Pharmacy, 3: 39–46. doi: 10.1016/j.scp.2016.03.001.

[37] Liu, S., Zhang, M., Kong, J., Li, H. and He, C. (2023). Flexible, durable, green thermoelectric composite fabrics for textile-based wearable energy harvesting and self-powered sensing. Composites Science and Technology, 243: 110245. doi: 10.1016/j.compscitech.2023.110245.

[38] Attia, N. F., Ahmed, H. E., El Ebissy, A. A. and El Ashery, S. E. A. (2022). Green and novel approach for enhancing flame retardancy, UV protection and mechanical properties of fabrics utilized in historical textile fabrics conservation. Progress in Organic Coatings, 166: 106822. doi: 10.1016/j.porgcoat.2022.106822.

[39] Raj, A., Chowdhury, A. and Ali, W. (2022). Green chemistry: its opportunities and challenges in colouration and chemical finishing of textiles. Sustainable Chemistry and Pharmacy, 27: 100689. doi: 10.1016/j.scp.2022.100689.

Chapter 9

Green Automation in our Family Life

9.1 Individual contribution to a greener world

The degradation of the environment has been induced by humans, and we, humans, will have to find solutions that allow the planet to continue with the necessary survival conditions for its existence. If industrialization has promoted a more pronounced deterioration of environmental sustainability, it is also true that the small acts of our lives can define whether we are available to contribute to a better environment, or not.

The rate at which the population of large cities produces garbage is frightening. Much of this garbage concerns packaging. There has been a growing concern with the materials used in these packages, which, most of them, can be made of recyclable and already recycled materials. Paper has been gradually replacing plastic, and producers have shown a growing concern to reduce the amount of material used. However, this concern is not generalized and the existing regulations in most countries, mainly in developing countries, but also in developed countries, is still not sufficiently persuasive, leading to a lower consumption of materials and a more adequate selection of the materials used. As, the reader has already seen that in hotels shampoo can be dispensed to the guest in a fixed rechargeable system, or in small individual doses. The impact on the environment is clearly different between these two solutions, but it seems that there is still no regulation that imposes the use of refillable systems to the detriment of the proliferation of packages with individual doses, which generate an enormous amount of waste, when compared to rechargeable systems. This is just a small example of solutions that are already implemented, but not widespread. The non-generalization of the good practices is contributing to environmental degradation and will probably only be considered when hotel companies are required to comply with certain regulations, as there seems to be no vision and willingness to act in small, but important details. Consumer awareness is also a matter that needs to be largely taken care of, because if the consumer complained about the way shampoo is dispensed to him/her by the hotel, the hotel's top management would probably have already changed its way of dispensing shampoo to its guests. It is precisely the joint carelessness of product suppliers and consumers that has led to the chaotic environmental situation we are currently experiencing, and which we will have extreme difficulty in getting out of. And this is just a small example of how we can contribute to reducing the waste produced, but this example can be multiplied by the countless acts that we practice every day, allowing us to realize that we carry out small actions during each of the days of our lives that, if are not properly cared for, have a significant impact on environmental conditions, and that the "punishment" that nature is giving us is the result of all those little thoughtless acts that we have been practising.

According to data reported in the work by Cappelletti et al. [1], about 88 million tons of waste are produced in the European Union per year, and about 53% of this volume corresponds to household waste. It should also be noted that around 45% of this waste is generated due to incorrect consumers' behavior or lack of awareness in the way they generate this waste, which can be avoided. Some of that garbage corresponds to food, either through leftover meals or through the expiry date of products. This problem is quite common in families where stock management is not properly cared for. However, automation can be a valuable aid in managing which meals to cook depending on the stock and expiry dates of the products, which quantities to produce to avoid wasting food, managing the acquisition of new food products, etc. To this end, Cappelletti et al. [1] associated an intelligent freezer, capable of providing information about the existence of products, to a computer application and a set of procedures capable of guiding the user in the activities of cooking meals, including issuing lists of needs. The developed tool was particularly useful in reducing product waste due to the expiry date, i.e., in stock management, considering the expiry date of each product. However, the tool can be greatly improved, as it could allow including recipes that optimize stocks, reduce quantities to what is usually needed, and even suggest recipes that are best suited to the health status of users. In this regard and starting from the most complex part which would be the control of stocks in the freezer, and which is provided by the hardware that equips the container, there is potential for the software to be greatly improved and the environmental and social sustainability to be improved, reducing the production of garbage, the waste of food, and increasing the health and wellbeing of users.

In addition to the waste produced, there are countless factors in our daily lives that are implicitly linked to the environment, and one of them is consumption, in its most diverse aspects. Within the issues related to consumption, automation can create the necessary conditions for saving the environment in some important factors, such as water consumption and energy consumption. In addition to other aspects connected with the construction of the buildings themselves, which will be dealt with later, buildings can provide conditions that, in an almost mandatory way, induce a reduction in water consumption [2] and, mainly, energy [3]. Automation has significantly advanced in smart home automation, allowing to correct careless behavior, taking care of the environment.

9.1.1 Reducing the water consumption at home

The reduction of domestic water consumption can be achieved through automation in several ways:

- through measurement systems that clearly show water consumption to the user, creating increased mental pressure that induces the desired savings;
- creation of heating systems closer to the place where hot water is consumed, avoiding waste, and increased cold water until water is supplied at the desired temperature;
- controlled irrigation of gardens and green areas;
- recirculation and purification of water used in swimming pools;
- use of rainwater for subsequent irrigation or filtering and treatment for use as drinking water;
- among others.

Raising consumer awareness regarding the water they consume is extremely easy to do and, depending on the country and the consumption understood as the minimum necessary amount of water per inhabitant, automation could create alert systems and financial or other incentives that reward the reduction of the water consumption, given this is a good that tends to become scarce. The display of periodic water consumption and alert systems based on recommended consumption would help to increase citizens' awareness, contributing positively to water reduction [4].

Hot water heating systems in homes are not always correctly optimized, placing the heating systems close to the preferred place of consumption. The waste of cold water that occurs until the system supplies hot water is partly unavoidable, due to the cold water accumulated in the pipes between heating and consumption points, but a careful design of the installations could allow a significant reduction of this waste. In this case, automation does not play such a significant role, as it essentially depends on the care of the designer, or on the existing regulation, when it exists, that conditions the designer to adopt heating strategies, placing them close to the places of consumption. The modeling of population water consumption and the use and programming of water heating systems using solar energy has also been studied, and it was found that the use of these panels can also contribute to the reduction of domestic water consumption [5].

Garden areas and backyards are places where water consumption is mandatory, and usually abundant. However, this watering, despite being necessary, should be carried out judiciously. Watering criteria depend on weather conditions, as well as the soil and species that need to be watered [6]. Once these criteria have been established, irrigation can be carried out on a basis of minimum consumption, or it may not have any criteria and be carried out in an excessively abundant manner, not considering the real needs. Automation, duly controlled by decision support systems based on algorithms that properly read meteorological conditions and adapt consumption needs to those conditions, considering the species that need to be watered and the state of the soil, will allow for regulated consumption of water, which will limit consumption to only what is strictly necessary, considering environmental sustainability [7]. The use of drip irrigation systems or the timing of irrigation systems and adaptation to the hours at which it should be carried out, is something extremely simple to carry out, and which brings significant environmental and economic benefits. Both public and private gardens, as well as backyards and small farms, are still in a very early stage of applying systems of this kind, mainly in developing or less developed countries. Thus, there is still a wide field of progression for the design and implementation of these systems based on sensors and actuators, with a centralized computerized command.

Private pools at homes, as well as public pools, are also a problem for excessive water consumption. However, its recirculation and treatment can be a reality, which is not yet properly explored, mainly at the domestic level. Automation can play an equally important role in this regard. The analysis of the state of the water is easily carried out through electronic systems that promote this analysis and can send a signal to a centralized system that promotes the recirculation and treatment of the water until it reaches the levels of purity necessary for healthy use of these pools [8]. Although the replacement of water in swimming pools is not carried out daily, consumption is quite high for each emptying and filling of the same, which translates into a high overall consumption, and which can be significantly reduced, providing it is properly controlled by automatic systems. The health guarantee will be assured, and water consumption can be significantly reduced [9].

In most cases, rainwater is not conveniently used for day-to-day activities, except for irrigation purposes. However, in recent years, studies are emerging with a view to taking advantage of this resource, using the collection and treatment of rainwater for domestic purposes [10-12]. Obviously, this use needs space and associated technology, in the sense of storing water in an appropriate place, also filtering, and treating it, enhancing, and increasing the possibilities of its use in any type of domestic activity, and even as drinking water. Technology and automation can play a key role in this rainwater harvesting cycle, through managing its collection, analysis, filtering, and water treatment. As collected water usually do not have a uniform composition over time, it will need that diversified actions are taken and with different degrees of intensity, depending on its quality.

It can thus be seen that automation is already considered an almost mandatory way for the proper use and management of water in its most diverse aspects. The previously mentioned systems imply adequate control in the analysis, treatment and management of the water, which is only possible through computerized systems and a strong component of sensors and actuators. Depending on the data collected, which can be extremely variable in time, they will analyze the conditions and act accordingly, guaranteeing the adequate and expected uniformity in the quality of the supplied water. With the tools indicated above, it becomes possible to carry out automatic water management, which will certainly contribute to a more rational consumption of water. This automation will make water management easier for all those who are more careless with this aspect and will make life easier for everyone else who already has concerns about this precious natural resource. Technology implies investment. The governments of more developed countries, given the scarcity of water that will be felt in several densely populated areas, will certainly promote policies to support the management and reuse of water, with a view to improving the sustainability of the planet [13].

9.1.2 Reducing the power energy consumption at home

If water is a natural good that we all have an obligation to consume responsibly, electricity is a good that is almost as necessary as water, especially for developed and developing countries, but which needs to be produced. As mentioned earlier in this chapter, energy production can be generated through renewable resources, but a large part of this electrical power is still generated using materials that produce effluents that are harmful to the environment (coal and other hydrocarbons), or even dangerous to human health and even humans' existence (atomic energy). Reducing energy consumption has been a growing concern, an issue which took on particular prominence in 2022, due to geopolitical issues between Russia and Ukraine, leading to a significant increase in energy prices. However, efforts to rationalize energy consumption have already come a long way ago, having only been intensified due to the needs that emerged in the meantime.

Observing domestic energy consumption, it appears that the largest aspects of this consumption are related to water heating and building air conditioning. Consumption with lighting, cooking, conservation of food and entertainment, although important, do not have the same influence on the overall calculation of energy consumption in residential buildings as air conditioning and water heating. There is a new aspect of consumption that is related to recharging vehicle batteries, related to people's need for mobility. On the other hand, energy consumption has consequences for the environment, which are already perfectly identified. Lowering energy consumption in homes is a priority to which the governments of some countries have given special importance. In this group are mainly the countries most affected by the cold, where the insulation of buildings is a critical factor. However, insulation is part of what is called a passive strategy, i.e., after the building has been built or renovated, residents need not do anything else to save energy. In recent years this strategy has been reinforced with immense studies that aim to make buildings self-sufficient in terms of energy, mainly through passive strategies, but also through active strategies [14]. Irulegi et al. [14] developed a housing prototype that mixed passive and active strategies, essentially taking advantage of solar energy, to create a completely self-sustaining housing in energy terms, but which guaranteed comfort and an adequate sense of wellbeing to its occupants. The concept integrated architectural solutions and technology, also taking advantage of natural lighting, and making use essentially of natural materials conveniently aligned with the initially established self-sustainability purposes. In this prototype, an intelligent passive strategy was also used, placing photovoltaic panels as a way of isolating the facade of the house from solar radiation, which promotes the generation of electricity and reduces overheating of the house, avoiding the need for forced cooling. The exterior panels of the remaining facades were also selected considering their thermal behavior corresponding to the sun exposure in which they were oriented. Passive strategies, in addition to adequate insulation considering the location where the building is located, also allow for the orientation of the building, the materials used in each area of the building, water heating using solar energy, and

Figure 9.1. General aspect of the Ekihouse prototype developed by Irulegi et al. [14].

the generation of electricity through photovoltaic panels. So that the energy balance between consumption and generation can be a reality, and so that the comfort of the inhabitants is preserved, some active techniques can also be used, depending on the direct action of the inhabitants, which can contribute essentially to reduce energy consumption.

Technology assumes special prominence in projects to improve the sustainability of buildings and reduce energy consumption, as it combines a whole set of sensors, actuators, and control systems that are indispensable for reading, intervening, and managing the different active elements that promote these savings. Del Rio et al. [15] made an integrated analysis of how the culture of the inhabitants, energy savings and environmental sustainability are correlated in four different countries, analyzing their ability to adopt smart home technologies. In fact, any change in habits normally entails a set of drivers and barriers that need to be dissected, trying to mitigate some difficulties in the corresponding implementation. The countries considered in that study were the United States of America, Japan, the United Kingdom, and the United Arab Emirates, focusing on different aspects, such as social, technical, political, economic, and environmental. Although all the countries considered in the study have high average wages and there are no perfectly defined economic barriers to the acquisition/adoption of new technologies due to economic limitations, these countries have quite different socio-political and socio-cultural patterns, which conditions the adoption of these technologies in their homes.

The lifestyle, the populations' average age, weather conditions, the concept of privacy, owning or not the housing, language barriers, aptitude for new technologies, religious or gender limitations, environmental awareness, social awareness, among others, conditions the adoption of home automation in these countries. A luxurious lifestyle tends to increase the adoption of new technologies in housing, saving the inhabitants work, even if they do not have the necessary environmental sensitivity for the benefits they are inducing in the environment. On the contrary, populations that tend to live a simple life, having a strong passion for environmental sustainability, will have a positive influence on the adoption of home automation for their homes. On the other hand, age can be a strong constraint, as the older population has less enthusiasm for new technologies, who could may offer some resistance to the understanding and use of new computer applications responsible for managing systems. The fact that these systems are usually connected to powerful servers that monitor and update the systems, can create a feeling of insecurity in older users, or whose culture is less open to sharing information. The language barrier can also become a problem in certain countries, who are too deeply rooted in their way of expression and with a more distant contact with the most common languages in terms of technology and software. Although there has been a huge evolution in automatic translation into the most diverse languages, the form of expression is not always 100% compatible, creating some barriers, especially for older people or people with a lower educational level. Developers of computer applications are perfectly aware that these applications need to be as user-friendly as possible, so that familiarization with the application is quick and the user can take full advantage of all the features available to them. However, the ability to use computer applications varies from person to person, and older people tend to have greater difficulties in using these applications. In certain cases, it becomes necessary to use multiple platforms, depending on the systems and corresponding suppliers. This situation could be an obstacle to the adoption of solutions that allow effective energy savings, with the respective harmful consequences for the environment. Adherence to and practice of certain religious precepts, as well as gender, could also be the source of certain limitations in the adoption of new technologies linked to energy saving. Access to education and certain practices is not universally accepted, thus, there may be greater difficulty in accessing this type of technology by women in some countries. On the other hand, the educational level also conditions access to information and mental openness to the pursuit of environmental objectives. Social awareness, on the other hand, can act positively or negatively, depending on the society in which people are placed. The adoption of new energy saving technologies may be seen as a symbol of modernity and technological development, but it could also act negatively as it may represent a clear distinction of purchasing power, which could create revolt and envy in needier societies. The balance between driver and barriers should be clearly positive to encourage the adoption of increasingly effective energy saving practices.

On one hand, Li et al. [16] reported as the main positive factors in the adoption of home automation, the possibility of an efficient management of energy consumption, economic benefits, and an improved quality of life. On the other hand, factors such as a limited perception of the benefits of home automation and how it works, economic problems that impede investment, mistrust, and anxiety regarding the intensive use of technology, distrust regarding security, and harmful social influences are pointed out as an impediment to the adoption of home automation and evolution to this level of technological involvement. It should also be noted that some of the people surveyed pointed out as risks the lack of privacy, possible negligence in the consumption of electricity because the generation is domestic, and even the possibility of destroying domestic life.

The relatively common problem of difficulty or anxiety in handling computer applications on which home automation systems are based was also identified by Ruiz et al. [17], who developed a model capable of capturing the semantics of devices, services, needs, goals, and actions, with a view to future projects in this area that present as main characteristics being self-adaptive, self-evolutionary and possess critical capacity to regenerate. However, the authors felt that the work should go even further, so they defined new objectives, which included providing home automation systems to generate new automation rules covering a wide range of situations, monitoring the system to update the set of rules when any of the necessary resources are changed/updated, and to take part of all the collected information to increase the capacity of self-learning.

Among many other advantages, such as increased comfort and quality of life, the adoption of Smart Home Automation brings clear advantages in terms of energy savings. Louis et al. [18] monitored the application of this system when installed in a house considered to be representative of what a family home is in Finland, stating that the energy savings achieved significantly improve the environmental impact, through 13% reduction in CO_2 emissions that would be at the base of energy production necessary to feed this house. Part of this reduction was achieved through more adequate planning of the operation of devices that consume more energy, transferring their operation from peak hours and overlapping with other equipment, to times of lower energy consumption, which can easily be managed by the software responsible for managing the household's energy consumption, through the introduction of algorithms suited to the tariffs charged, energy consumption peaks at different times of the day, and the consumer's energy needs. In addition to promoting savings for homeowners, these solutions are also contributing to meeting the objectives set by the EU, which include a reduction of between 80 and 95% in green gas emissions, with reference to the year 1990 and the year 2050 as the deadline for meeting the objective (Energy Roadmap 2050), and this emission reduction strategy is not exclusive to Europe, although the strategy of some countries is not completely clear and evident. The emission reduction achieved in the study by Louis et al. [18] is encouraging and allows realizing the potential that the adoption of smart house automation can present considering the reduction of environmental damage that needs to be carried out worldwide and shows how much energy saving can contribute to this goal. In a later study by the same authors [19], a life cycle assessment was carried out considering two smart home automation approaches and it was found that the installation of smart meters pays off economically and environmentally, with payback times of 3.5 mon and 11 mon, respectively, but the installation of smart plugs does not pay off economically, bearing in mind the period considered for the study (5 yr).

Van Thillo et al. [20] also carried out a study on energy savings achieved through the implementation of smart home automation systems, concluding that there are different values and opinions in literature, making it difficult to realize the real benefits brought by the adoption of these systems. While San Martin et al. [21] referred to energy savings of about 26% with the implementation of smart home automation, moving from class C to class B, Firlag et al. [22] referred to gains of only 13%, although the basis for comparison is not the same. In fact, it will always be difficult to establish comparisons and replicate results, given that each building has its own characteristics and that boundary conditions are never the same. It should also be mentioned that the characteristics of the building itself may include three aspects: (a) the architecture and positioning of the building; (b) the characteristics of the smart home automation installation; and (c) the installation contextual parameters. In the case of the study by Van Thillo et al. [20], four different aspects were considered, namely heating, hot water supply, lighting and shading of some areas of the house. The study provided different savings values for each of the aspects considered, with more significant savings being obtained in housing heating, going from class D to class A. In lighting, savings are more modest, but even so it is possible to move from class C to class A. In shading, despite moving from class D or C to class B, the savings are still encouraging, although not as relevant as in heating. That work [20] also presented a detailed study of which factors can contribute in a more accentuated way to the mentioned savings, which could help to focus attention on the factors that effectively have greater advantage in improving energy savings when implementing a smart home automation system. As usual, the simulations contain assumptions and simplifications, so the results described by Van Thillo et al. [20] have the usual associated error.

The need for energy optimization has also led to the development of simulators that make it possible to draw different scenarios in view of a given consumption framework, and subsequently proceed with their adjustment according to the user's profile [23]. The algorithms developed in the meantime can be validated and improved with a view to promoting the desired reduction in energy consumption. However, to achieve that, it is necessary that the systems can collect and processing the necessary and desirable amount of data in real time, through various sensors and consumption analyzers [24]. Despite several models being developed capable of creating datasets that allow adjusting the operation of certain home automation systems to the user's profile in a first approach considering the user's consumption history, models based on real-time data collection and interaction are more favorable. However, based on the history of other users, it is possible to define a previous model that best fits each user, considering account the number and sequence of events that usually take place in a given dwelling, as well as the probability and frequency of occurring. However, the optimization usually requires a further adjustment based on real-time data. Thus, interactive systems can be created that, by collecting data in real-time, allow the control adjustment of the algorithm parameters, improving efficiency in energy management. However, simulators should consider climate variations, which is uncommon in simulators created in the meantime [25]. The use of sensors to monitor the temperature and humidity of the air could provide crucial data to the system, which, crossed with the considered dataset, could provide much more reliable simulation results, contributing to an even greater energy efficiency of these systems.

The systems can also provide users with energy saving recommendations, trying to contribute to changing the habits of a given family, considering the family's behavior and consumption patterns that the system registers. The recommendations aim to encourage inhabitants to change some consumption patterns, namely to move certain energy-consuming activities out of peak consumption hours, detect energy-consuming activities that can be avoided without losing the comfort and safety conditions of the house, reschedule certain activities to avoid overlapping consumption by different devices, as well as use IoT to help residents remotely control the activation or shutdown of certain devices when they are away from home. With this aid, users will achieve much more efficient energy management, with the consequent saving of economic resources and burden for the environment, without this representing an additional concern or jeopardizing their quality of life and wellbeing. Some household appliances have a usual duty cycle, which is known to users. This cycle can be studied and framed in off-peak hours, allowing tasks to be carried out in a timely manner and with the necessary comfort and safety, without an exaggerated overlap of consumption load, and using more favorable schedules. After a period of learning the system, considering consumption habits and data that favor more efficient energy consumption, and knowing the operating standards of each of the devices connected to the system in each home, the smart home system can suggest or even manage the sequence and hours of the different devices operation, allowing consumption to be harmonized according to the most economically and environmentally favorable consumption pattern.

Just as the decision to adopt smart house automation systems can be affected by several factors related to the family owning each house, there are also factors that underlie the adoption of this technology by hotels and other accommodation sites, in terms of tourism. Saving energy in this business area is of vital importance, not only for the environmental part, which is the main focus of this work, but also for the economic part that sustains the business. However, when it comes to making strategic decisions for a diverse public not known beforehand, it is necessary to take risks and focus on certain market segments, based on a relatively deep knowledge of what are the expectations of customers in each market range, and for which market ranges they should work. When the market range with which it is intended to work is technologically advanced, they value the technological progress, and possibly also value the environmental component that may be associated with it. Thus, the risks of innovating in the form of business will be lower, and hotel and accommodation owners in general could consider implementing innovative smart automation systems in tourism facilities, without running the risk of going against customer expectations. There are some factors that are associated with this type of decision, namely the ability to innovate and the degree of innovation to be implemented, customer demand and expectations. Moreover, the impact that such decisions could have on the reduction of jobs should also be considered. These factors were properly structured and dissected by Tussyadiah et al. [26], describing the implications that decision-making considering technological evolution could have on demand. However, it is evident that, more than a market option, it is also a conscience option, considering the environmental impact it will cause. Moreover, it will certainly be an option that will lead to certain market fringes feeling that this type of offer better adapts to their environmental awareness, having the sensation that this also corresponds to contributing to a better environment.

9.1.3 Smart home automation

The development of the X10 platform gave rise to what would come to be called smart home automation, still in the 1970s. This platform sent batches of information through existing electrical cabling in residential buildings [27]. However, only more than a decade later, the subject was revisited again, when the City-Place building was designed, incorporating technology that allowed an integrated control of lighting systems, air conditioning, elevators and other equipment, through a computerized system. This system allowed the management of communications and email, also applying intelligent materials [28]. This constituted the turning point in relation to the development of the concept of "smart buildings", through the passage into practice of the concepts previously developed regarding the integration of information. Starting from the concept of information integration, until understanding everything in which automation can help to increase the comfort and daily activities of families in their homes, it was a process of natural evolution, which has been taking place naturally over time, but which has essentially focused on three main areas: efficient energy management, safety of people and their assets, and health care [29].

Smart home automation is something that is starting to become more and more frequent in the design of new buildings, mainly to the drastic reduction in the price of most of the necessary automation devices, the growing development of applications that allow a much more effective management of most of the domestic activities related to the needs of users and their comfort, as well as a growing awareness of all the benefits that can be derived from the use of this technology by consumers. This creates additional pressure on builders to provide these solutions in new buildings, and even in the recovery of old buildings. Smart home technologies allow users to be connected to various devices in their home and control them directly, or following previously defined algorithms, improving their efficiency. The capacity of devices capable of being connected and controlled remotely has grown significantly, so it is possible to time, program, anticipate the operation of any device, or remotely turn off devices, making it easier to manage the daily lives of users and rationalizing the energy or water consumption and functioning of these devices. This more effective control translates into functionality benefits for users, and can also represent significant gains in environmental terms, due to savings in energy consumption, as seen previously, and even in stock management, waste generation and water consumption. The devices can be controlled through wired or wireless systems and, through the computer applications developed for the purpose. In addition to control and programming, the user can also have some tools that allow saving and processing usage data. With this, users can manage devices more efficiently, analyzing how certain decisions can result in savings, which not only translate into economic benefits, but also into benefits for the environment. This duality of benefits increases users' motivation for adopting these technologies, as certain consumers may have a greater focus on saving costs, while others will tend to understand more clearly how their options are benefiting the environment.

The adoption of wired or wireless controlled systems is still an issue that divides consumers who adopt these technologies. Wired technology allows for greater security, although it brings inconvenience in terms of installation, especially when it comes to the recovery of older buildings, where it is necessary to provide infrastructure for the entire building. Furthermore, the configuration and integration of different devices may require the intervention of specialized technicians, due to the possible existence of communication or connection incompatibilities. Wireless technology is usually associated with a greater security risk, although it has a greater interconnection capacity, making the connection of devices from different sources/manufacturers quite easy for the user, which can serve as a stimulus to its adoption.

The adoption of a home automation system in the case of urban buildings requires a prior analysis that weighs the costs and the respective benefits, considering factors as diverse as the population that will probably inhabit the building, their level of education technology, the way they want to take advantage of the building, the environmental conditions in which it is inserted, the expected degree of comfort, etc. Buildings placed in areas where the population that will acquire and inhabit them is older and less prone to technology, will tend to look for less technologically developed buildings. On the other hand, buildings in areas intended for a younger population, at the beginning of their careers, but with a higher technological education, more ingrained in environmental concerns, and more demanding regarding technological needs, will have to be provided with another type of comfort, more adequate to the real expectations of buyers and inhabitants. Aste et al. [30] developed a framework capable of helping in deciding the level of automation to be applied in each building, which intended to integrate data from statistical models with simulation models, and understand to what extent the intelligent energy management of a given building can be an advantageous tool factor for those who build it and for those who will inhabit it. These frameworks should consider the estimated lifetime of the building, balancing the positive and negative factors of implementing a system like this, as well. To this end, the models will have to consider energy consumption, the corresponding emissions, the comfort provided by the system and its cost, in addition to the other factors mentioned above. These systems also require the collection and processing of data generated and recorded by the system in the meantime, sometimes implying corrections to the algorithms programmed over time, with a view to a better adjustment to the actual conditions of use. In addition, uniformity and transparency in the data collected and recorded should be ensured, which will provide a better analysis of other similar situations in the future. This information will increase the know-how of the companies that supply these systems, making them better able to advise on the system that best fits to each case and each environment. In addition, this data can be treated statistically, making later estimates much more reliable, making the application criteria more robust in the design phase. This will allow the systems to be optimized and, with that, the energy savings to be more effective. It is also necessary to correctly monitor the system throughout the life of the building, in order to detect possible deviations from the initial estimates. Alert systems can be created that, in the event of a deviation greater than the stipulated, issue notifications able to generate corrective actions for the system, so that it remains within the standards initially established.

As tariffs are dynamic and depend on the periods of greatest overall consumption on the power grid, the management of the source where energy is drawn at each moment also needs to be controlled. This factor is even more critical when the home has its own means of storing energy. Thus, given the tariff in use, the existence of an algorithm that managed where to get the energy at each moment would be extremely useful. In fact, at a given time, even with energy stored in the home, it could be more profitable to consume energy from the power grid itself while maintaining stored energy levels, as this consumption is situated within an economic tariff period, and energy is accumulated for periods consumption and subject to higher tariffs. However, to do so, it is necessary to consider a history of consumption and analyze at each moment which option is the most sensible. Considering only the management of each dwelling, Chauhan et al. [31] developed an algorithm that taking into account the cost of energy at each moment, the comfort of the inhabitants and the perspectives of consumption in the short term, allowing consumption to be directed to the power grid or to the energy storage system based on these factors. Although the main purpose of the developed algorithm is essentially reducing the customer bill, this intelligent management is not only beneficial for the consumer, but also for the power grid, as it tends to avoid network overloads at peak consumption times, and makes energy storage profitable when the network is less requested, promoting the charging of the energy storage system. The large-scale application of these intelligent systems promotes a smoothing of consumption peaks, distributing them over longer periods, which largely benefits everyone involved.

Real-time information on energy consumption can be an extremely beneficial factor for controlling the consumption of each home. Effectively, this information allows inhabitants to be more aware of the consumption they are using, being able to act immediately on it, adjusting the need at the time, or taking into account other factors such as the cost of energy in that time interval. Zorrilla and Ibrain [32] developed an intelligent real-time consumption information system, which allows both end users and energy distribution companies to be aware of current consumption, thus being able to make more responsible use of the energy resources. The system they developed has the advantage of being economical and not intrusive. In addition, the system allows to trace consumption patterns, which can be adjusted to factors such as the price of energy in each period of the day, thus allowing to switch on or off certain devices accordingly and optimize consumption. For the final consumer it can provide very important information regarding their daily routines, which can eventually be adjusted depending on the cost of energy at any given time, and for the energy distributors it allows knowing consumption patterns and predicting peak periods and consumption slowdown, adapting their purchase of energy according to demand at each time of the day, and establishing differentiated prices, so that customers are induced to change their consumption habits depending on the cost of energy.

The integration of smart home-related systems needs to be based on communication platforms that, preferably, can communicate with each other [33]. When such a system is installed in a building, there is a concern that it be designed so that all devices have easy communication with each other. However, there is still no standardized platform that can be universally adopted, allowing to ensure that communication between buildings, or between buildings and control entities (condominium management, or even municipal or national entities) is facilitated. This standardization of data availability is extremely important to ensure proper management of buildings over time, and so that data can be monitored by competent authorities, with a view to ensuring that buildings are actively contributing to the desired savings of energy resources. Although smart home automation has already been implemented for more than two decades, development has essentially been carried out based on the know-how of each player in the market and, in certain circumstances, commercially, it is convenient that the systems are not standard, thus helping to retain users. However, there is an urgent need to standardize the communication and data organization systems, which will allow for a more sustained technology development over time. In fact, buildings are built with a very long-life expectancy, and technology is developed at a brutal pace, which makes systems that are still in their initial deployment phase obsolete. The creation of a common communication base would allow, regardless of the equipment brand, the communication between the different devices among themselves, and between them and the monitoring and data collection systems to be much more effective, and not run the risk of becoming outdated in the short term. This standardization largely depends on the regulatory bodies, usually associated with the country's government, or a group of countries, forcing manufacturers to develop systems that follow a common standard, facilitating communication today, but also ensuring that future devices are also aligned by the same standards, preventing systems from becoming obsolete in the short term. In addition to common sense, the adoption of this type of standards helps to reduce the waste of resources, increasing the useful life of the systems. This also greatly facilitates the maintenance of the systems, preventing that, due to the failure of a device, an entire system needs to be abandoned, due to the incompatibility of the new devices with the existing equipment. In this way, it will be possible to contribute to a much more sustainable future, in which there is no waste of resources due to the lack of standardization of devices, in terms of communication and compatibility.

Given the severe need to save resources and minimize environmental degradation, smart home automation is something unavoidable. Many countries and communities are already funding programs to support the development of more sustainable energy solutions, a situation that turned out to be even more evident in the energy crisis experienced in 2022 and 2023, due to Russia's aggression against Ukraine, with one of the main focuses are the places of energy generation, as well as the reduction or cut of hydrocarbons mainly provided to Europe. This situation has awakened some politicians to the high energy dependence of more unstable countries, but which hold strong reserves of hydrocarbons. It was easily perceived that energy dependence could be a factor of important political influence, a situation that is highly unfavorable to countries that do not have these reserves. Thus, a more localized production of energy and the adoption of energy saving strategies are the best way to reduce or eliminate this dependence.

9.2 Sustainable construction

The construction sector can act in different ways to improve the environmental burden inherent in the construction and operation of commercial or residential buildings. Ideally, buildings should carry a neutral load on the environment, inter alia, by reducing the energy consumed. Technological development has permitted developments that allow, even today, a strong reduction in energy consumption in different ways, namely: (a) energy generation using the building's own resources, through the use of photovoltaic panels, energy generation through the use of wind (wind energy), energy generation through fuel cells, and reuse of waste for energy generation; (b) storage of this energy in batteries and supercapacitors; (c) use of solar collectors to heat water which can then be fed into a conventional heat pump [34].

Sustainable construction has deserved particular attention from structures such as the European Community for a long time, which launched a program called LIFE in 1992, through which it promoted practices able to improve the construction sustainability within the common European space. Fetsis [35] carried out a synthesis work on everything that this program brought to Europe in terms of innovation around sustainable construction. In fact, the program co-financed several projects with well-defined objectives: to reduce carbon emissions related to the construction sector and reduce energy consumption, with a view to mitigating climate change. Thus, the program supported the generation of renewable energy and projects related to improving energy efficiency. The reuse of materials was also another focus, increasing recycling and the use of recyclable materials, as this was one of the main problems encountered in traditional construction. However, automation also had a prominent place, through the development of temperature management and control systems, both in heating and cooling buildings. However, sector professionals also need to know the materials and technologies available to them, as well as the most appropriate construction techniques for each type of material and construction. Thus, the program also set aside funds to support awareness-raising and training actions for new, more sustainable materials, as well as innovative techniques for working and applying these materials, making workers aware of the importance of using these materials and construction techniques, clearly indicating the advantages and benefits for the environment that they can provide. This training and awareness-raising should take on a permanent character over time, given that the development of new materials and construction techniques is extremely active, and it also becomes necessary to be increasingly effective and efficient in the construction sector, given that each building built now, will have a long-life cycle, so it must be designed with that in mind. The renovation and retrofitting of buildings are another extremely important area on which the program was also focused, as it is possible to intervene in these buildings so as to improve their sustainability. Each renovation should consider all the possibilities to make buildings more energy efficient and incorporate the degree of automation that best suits the requirements of their inhabitants. The techniques and materials used in renovation and retrofitting are sometimes quite different from those normally used in new construction, so training should be specific to each area.

It was precisely in villa retrofitting operations that Ascione et al. [36] worked with a view to optimizing the energy consumption of recovered houses, and to analyze two different construction techniques, with a view to understanding which one had the best sustainability index: a lightweight house in reinforced concrete and a massive tuff-made villa, located in the Mediterranean coast (Italy and Greece). Regarding energy optimization, applying the Pareto's approach, they concluded that the best results could be achieved by combining transient energy simulations and a genetic algorithm. The energy study considered factors such as sun exposure, thermal insulation, possible reflectance, area of windows, etc. Consideration was also given to the introduction of renewable energies through solar panels, and the existence of air conditioning to adapt thermal comfort conditions. Considering the two case studies on the Mediterranean coast, the authors concluded that following the guidelines outlined by them, it would be possible to obtain energy savings of around 125 kWh/m^2, and that by adopting the best construction strategy, it would be possible to save up to around €140/m^2. According to the same authors, considering the usual climate in those regions, the solution would be to cover the entire roof with solar panels, including thermal insulation and reducing the glazed surface of the windows.

However, automation can also be extremely useful in using new materials, which in turn can be much more sustainable. In the last decade, with the development of various additive manufacturing technologies, the construction sector realized that it could also become an added value in the construction of buildings, since it is a highly automated process. Through this process, the buildings are built layer by layer, resulting in a practically monolithic construction, with very little human intervention. This type of technology also allows the use of a wide range of materials, which include more sustainable materials than those traditionally used in conventional construction. Raza and Zhong [37] presented a work in which they explored the use of geopolymers in the construction of buildings through additive manufacturing, using geopolymers consisting of FA, GGBS, SF and metakaolin resulting from industrial wastes and different 3D printing techniques. According to these authors, these materials have a high potential for use in the sustainable construction of buildings in the very near future. However, the application of this technology and these materials still has some limitations. In fact, the cost of geopolymers is still considerable, which constitutes an obstacle to a more generalized application of these materials. Furthermore, the binding agents (alkali activators) used are still produced on a relatively small scale, which leads to higher costs for the product, clearly noticeable in the final product. The local demand for materials that satisfy the necessary requirements for construction (geopolymers) is also a major challenge, as it will imply the creation of infrastructures capable of collecting industrial waste of interest for construction, which in turn implies its own logistics and a screening and classification process able to fit the quality requirements demanded by this type of construction. On the other hand, not using or reducing the amount of alkali activators applied leads to very long curing times between layers, which leads to very long construction times, making the process unattractive. The periodic use of steel reinforcements, as well as the search for geopolymers in the vicinity of construction sites are factors to be considered in future research work. It seems that, being a future possibility with high potential, there is still a long way to go so as to assess the quality of the materials and the economic viability of construction, constituting, however, an excellent alternative in terms of environmental sustainability.

With a view to ensuring maximum comfort for the inhabitants and principles of environmental sustainability, the Eco-Label was created. This concept intends to integrate construction details, day-to-day management, and the possible disposal of building materials, throughout the life cycle of the building [38]. The Eco-Label allows, in a standardized way, to inform the possible customer about the comfort and sustainability conditions of each house, allowing the buyer to take a decision based on a comparable basis between buildings, i.e., on measurable and analyzable factors in each building. In this way, the potential buyer will have a broader and more assertive notion of what he/she can find in the house in terms of comfort and respect for the environment. Basically, this is an adaptation to the construction market of something that is already common in many countries regarding the energy performance of domestic appliances. In this way, a comparative base is established based on the same analysis factors, which provides more reliable and sustained information to the potential buyer, passing on to him/her the decision to opt for almost sustainable solutions.

9.3 Sustainable and smart cities

The concept of smart communities has been widely dissected, considering the establishment of standards that correspond to what was understood to be called smart cities. Smart cities use systems and procedures able to improve local economy, transport, and traffic management, and contribute to their populations' wellbeing [39].

In fact, the smart-city concept brings together the environmental sustainability factors that the city intends to establish, with the wellbeing of its inhabitants [40]. For this strategy to work, adequate city planning is necessary. However, for this planning to be possible, it is necessary to establish which are the main drivers latent in the formation of a smart city. It was exactly this study that Nunes et al. [40] developed, establishing a set of guidelines and priorities that should be imposed, thus leading to the results that are intended to be obtained. The authors of that study took two case studies in completely different Portuguese areas in terms of location and climate, to validate their study, with the developed framework showing good adaptability to very different situations.

In a social aspect, cities should be structured so that populations can develop their lives in a way considered normal, balancing work with leisure and the pleasure of living. Work-related activities are increasingly intense and demanding, which induces in populations the desire to rest more effectively in their free time. Domestic activities, if simplified with the help of automation, will be a valuable aid for increasing the quality of life of the active population.

However, it would be extremely beneficial if the smart city concept could also have applicability in rural areas, where populations and their needs are significantly different from what are the standards of more developed cities [40]. Bogataj et al. [39] clearly referred to the loss of functional capacity of the population from birth to old age, listing in an objective way how automation can help improve the lives of these older people living in more remote villages, i.e., how automation can support the health of populations, provide information and training, and improve their domestic life. In this last aspect, in addition to the sensing that would issue alerts in the case of need for medical assistance, the monitoring and regulation of the ambient temperature, as well as the management of the delivery of medicines and food products, and assistance in daily routines, remote medical advice, and even control daily physical exercise programs, could significantly benefit the quality of life of populations and help them to remain more active and with a better quality of life. This help to avoid some desertification of areas that, due to emigration of younger populations, tend to be left behind and fail to actively contribute to the global economy.

However, automation can also be a valuable help in small towns where the older generations usually settle, and who need more care. Mainly in more developed countries, younger generations have a strong tendency to migrate to areas where there are greater and/or better job opportunities, while older populations tend to remain in the places where they were born or where they developed a large part of their active life. Some of these populations are even linked to agricultural activities, which retained or still retain them in more remote areas. However, due to their older age, they also need greater care, both in terms of medical assistance and even surveillance. Here, automation can also significantly help these populations to feel more supported, preventing the desertification of some areas due to the massive migration of the younger population.

Monitoring and triggering medical assistance can be one of the aspects in which automation can significantly help to increase the population's sense of security and wellbeing. Automation can also help to increase the quality of life of populations, reducing the workload they could be subject to, both in their active lives and in domestic lives, simplifying actions and procedures through home automation. The introduction of robotic systems for the execution of certain tasks carried out by these older populations, as well as the introduction of services based on the Internet of Things and Cloud Computing can, through adequate training of the populations, allow them to feel that there are better conditions to live in these places, and can also encourage the creation of younger jobs, dedicated to new technologies, which allows rejuvenating the population of these places.

References

[1] Cappelletti, F., Papetti, A., Rossi, M. and Germani, M. (2022). Smart strategies for household food waste management. Procedia Computer Science, 200: 887–895. doi: 10.1016/j.procs.2022.01.286.

[2] Sarabia-Escriva, E.-J., Soto-Francés, V.-M., Pinazo-Ojer, J.-M. and Acha, S. (2023). Economic and environmental analysis of domestic hot water systems for single-family homes, Energy and Buildings, 286: 112925. doi: 10.1016/j.enbuild.2023.112925.

[3] Bogin, D., Kissinger, M. and Erell, E. (2021). Comparison of domestic lifestyle energy consumption clustering approaches. Energy and Buildings, 253: 111537. doi: 10.1016/j.enbuild.2021.111537.

[4] Canale, L., Slott, B. P., Finsdóttir, S., Kildemoes, L. R. and Andersen, R. K. (2021). Do in-home displays affect end-user consumptions? A mixed method analysis of electricity, heating and water use in Danish apartments. Energy and Buildings, 246: 111094. doi: 10.1016/j.enbuild.2021.111094.

[5] Vengosh, A. and Weinthal, E. (2023). The water consumption reductions from home solar installation in the United States. Science of The Total Environment, 854: 158738. doi: 10.1016/j.scitotenv.2022.158738.

[6] Reyes-Paecke, S., Gironás, J., Melo, O., Vicuña, S. and Herrera, J. (2019). Irrigation of green spaces and residential gardens in a Mediterranean metropolis: Gaps and opportunities for climate change adaptation. Landscape and Urban Planning, 182: 34–43. doi: 10.1016/j.landurbplan.2018.10.006.

[7] Šiftová, J. (2021). Shaping the urban home garden: Socio-ecological forces in the management of private green spaces. Land Use Policy, 111: 105784. doi: 10.1016/j.landusepol.2021.105784.

[8] Chaves, R. S., Guerreiro, C. S., Cardoso, V. V., Benoliel, M. J. and Santos, M. M. (2019). Comparative Biochemistry and Physiology Part C: Toxicology & Pharmacology, 223: 53-61. doi: 10.1016/j.cbpc.2019.05.015.

[9] Reißmann, F. G., Schulze, E. and Albrecht, V. (2005). Application of a combined UF/RO system for the reuse of filter backwash water from treated swimming pool water, Desalination, 178(1–3): 41–49. doi: 10.1016/j.desal.2004.11.027.

[10] García-Montoya, M., Bocanegra-Martínez, A., Nápoles-Rivera, F., Serna-González, M., Ponce-Ortega, J. M. and El-Halwagi, M. M. (2015). Simultaneous design of water reusing and rainwater harvesting systems in a residential complex, Computers & Chemical Engineering, 76: 104–116. doi: 10.1016/j.compchemeng.2015.02.011.

[11] Leong, J. Y. C., Oh, K. S., Poh, P. E. and Chong, M. N. (2017). Prospects of hybrid rainwater-greywater decentralised system for water recycling and reuse: A review. Journal of Cleaner Production, 142, Part 4, 3014–3027. doi: 10.1016/j.jclepro.2016.10.167.

[12] Liu, X., Ren, Z., Ngo, H. H.,He, X., Desmond, P. and Ding, A. (2021). Membrane technology for rainwater treatment and reuse: A mini review. Water Cycle, 2: 51–63. doi: 10.1016/j.watcyc.2021.08.001.

[13] Wang, W., Wu, F., Yu, H. and Wang, X. (2023). Assessing the effectiveness of intervention policies for reclaimed water reuse in China considering multi-scenario simulations. Journal of Environmental Management, 335: 117519. doi: 10.1016/j.jenvman.2023.117519.

[14] Irulegi, O., Torres, L., Serra, A., Mendizabal, I. and Hernández, R. (2014). The Ekihouse: An energy self-sufficient house based on passive design strategies. Energy and Buildings 83(2014): 57–69. doi: 10.1016/j.enbuild.2014.03.077.

[15] del Rio, D. D. F., Sovacool, B. K. and Griffiths, S. (2021). Culture, energy and climate sustainability, and smart home technologies: A mixed methods comparison of four countries. Energy and Climate Change, 2: 100035. doi: 10.1016/j.egycc.2021.100035.

[16] Li, W., Yigitcanlar, T., Erol, I. and Liu, A. (2021). Motivations, barriers and risks of smart home adoption: From systematic literature review to conceptual framework. Energy Research & Social Science, 80: 102211. doi: 10.1016/j.erss.2021.102211.

[17] Ruiz, A. R., Alises, D. V., Navarro, R. C., Romero, M. J. S., Chaparro, J. D. and López, J. C. L. (2021). Leveraging commonsense reasoning towards a smarter Smart Home. Procedia Computer Science, 192: 666–675. doi: 10.1016/j.procs.2021.08.069.

[18] Louis, J.-N., Caló, A. and Pongrácz, E. (2014). Smart houses for energy efficiency and carbon dioxide emission reduction. ENERGY 2014: The Fourth International Conference on Smart Grids, Green Communications and IT Energy-aware Technologies, 44–50.

[19] Louis, J.-N., Caló, A., Leiviska, K. and Pongrácz, E. (2015). Environmental impacts and benefits of smart home automation: Life cycle assessment of home energy management system. IFAC PapersonLine, 48(1): 880and885. doi: 10.1016/j.ifacol.2015.05.158.

[20] Van Thillo, L., Verbeke, S. and Audenaert, A. (2022). The potential of building automation and control systems to lower the energy demand in residential buildings: A review of their performance and influencing parameters. Renewable and Sustainable Energy Reviews, 158: 112099. doi: 10.1016/j.rser.2022.112099.

[21] San Martin, J. P., Garcia-Alegre, M. C. and Guinea, D. (2017). Reducing thermal energy demand in residential buildings under Spanish climatic conditions: qualitative control strategies for massive shutter positioning. Build Simul, 10: 643–661. doi: 10.1007/s12273-017-0360-5.

[22] Firląg, S., Yazdanian, M., Curcija, C., Kohler, C., Vidanovic, S., Hart, R. and Czarnecki, S. (2015). Control algorithms for dynamic windows for residential buildings. Energy Build, 109: 157–173. doi: 10.1016/j.enbuild.2015.09.069.

[23] Fakhar, M. Z., Yalcin, E. and Bilge, A. (2023). A survey of smart home energy conservation techniques. Expert Systems with Applications, 213: 118974. doi: 10.1016/j.eswa.2022.118974.

[24] Synnott, J., Nugent, C. and Jeffers, P. (2015). Simulation of smart home activity datasets. Sensors, 15: 14162–14179. doi: 10.3390/s150614162.

[25] Alsalemi, A., Ramadan, M., Bensaali, F., Amira, A., Sardianos, C., Varlamis, I., et al. (2019). Endorsing domestic energy saving behavior using micro-moment classification. Applied Energy, 250: 1302–1311. doi: 10.1016/j.apenergy.2019.05.089.

[26] Tussyadiah, I. P., Tuomi, A., Ling, E. C., Miller, G. and Lee, G. (2022). Drivers of organizational adoption of automation. Annals of Tourism Research, 93: 103308. doi: 10.1016/j.annals.2021.103308.
[27] Kravchenko, Y., Starkova, O., Herasymenko, K. and Kharchenko, A. (2017). Technology analysis for smart home implementation. In: 2017 4th International Scientific Practical Conference Problems of Infocommunications. Science and Technology (PIC S&T), pp. 579–584.
[28] Omar, O. (2018). Intelligent building, definitions, factors and evaluation criteria of selection. Alexandria Engineering Journal, 57(4): 2903–2910. doi: 10.1016/j.aej.2018.07.004.
[29] Marikyan, D., Papagiannidis, S. and Alamanos, E. (2019). A systematic review of the smart home literature: a user perspective. Technological Forecasting and Social Change, 138: 139–154. doi: 10.1016/j.techfore.2018.08.015.
[30] Aste, N., Manfren, M. and Marenzi, G. (2017). Building automation and control systems and performance optimization: A framework for analysis. Renewable and Sustainable Energy Reviews, 75: 313–330. doi: 10.1016/j.rser.2016.10.072.
[31] Chauhan, R. K. and Chauhan, K. (2019). Building automation system for grid-connected home to optimize energy consumption and electricity bill. Journal of Building Engineering, 21: 409–420. doi: 10.1016/j.jobe.2018.10.032.
[32] Zorrilla, M. and Ibrain, A. (2019). Bernard, an energy intelligent system for raising residential users awareness. Computers & Industrial Engineering, 135: 492–499. doi: 10.1016/j.cie.2019.06.040.
[33] Toschi, G. M., Campos, L. B. and Cugnasca, C. E. (2017). Home automation networks: A survey. Computer Standards & Interfaces, 50(2017) 42–54. doi: 10.1016/j.csi.2016.08.008.
[34] Lluna, A., Benítez, I., Monreal, J. and Díaz, I. (2010). Towards zero energy balance in tertiary buildings. IFAC Proceedings Volumes, 43(1): 321–326. doi: 10.3182/20100329-3-PT-3006.00058.
[35] Fetsis, P. (2017). The LIFE Programme–Over 20 Years Improving Sustainability in the Built Environment in the EU. Procedia Environmental Sciences, 38: 913–918. doi: 10.1016/j.proenv.2017.03.179.
[36] Ascione, F., Bianco, N., Mauro, G. M. and Napolitano, D. F. (2019). Retrofit of villas on Mediterranean coastlines: Pareto optimization with a view to energy-efficiency and cost-effectiveness. Applied Energy, 254: 113705. doi: 10.1016/j.apenergy.2019.113705.
[37] Raza, M. H. and Zhong, R. Y. (2022). A sustainable roadmap for additive manufacturing using geopolymers in construction industry. Resources, Conservation & Recycling, 186: 106592. doi: 10.1016/j.resconrec.2022.106592.
[38] Franzitta, V., La Gennusa, M., Peri, G., Rizzo, G. and Scaccianoce, G. (2011). Toward a European Eco-label brand for residential buildings: Holistic or by-components approaches? Energy, 36: 1884–1892. doi: 10.1016/j.energy.2010.09.021.
[39] Bogataj, D., Bolarin, F. C., Kavšek, M. and Rogelj, V. (2020). Smart silver villages as part of social infrastructure for older adults in rural areas, IFAC PapersOnLine, 53(2): 16914–16919. doi: 10.1016/j.ifacol.2020.12.1233.
[40] Ismagilova, E., Hughes, L., Dwivedi, Y. K. and Raman, K. R. (2019). Smart cities: Advances in research—An information systems perspective. International Journal of Information Management, 47: 88–100. doi: 10.1016/j.ijinfomgt.2019.01.004.

CHAPTER 10

Case Studies

10.1 An overview about some case studies in making processes greener using automation

Consumers who are more informed about environmental issues have begun to exert some pressure to understand what efforts companies are making, and which ones they intend to make in the future, with a view to reducing the environmental impact of their industrial activity. Despite this pressure still being manifestly insufficiently in view of the needs, the most attentive companies are already advertising more sustainable products, trying in this way to capture consumers who more easily adhere to these types of arguments. However, these arguments are only valid in environmentally enlightened societies, and this is essentially the case in developed countries. In the United States, Levi's advertises more sustainable pants, because it changed the finishing process for jeans, creating a slogan that tries to take advantage of more environmentalist consciences: 'water<less'. Innovation in this area is extremely useful, given the jeans sold by the brand and the amount of water consumed, as well as the announced reduction, which can reach 96% in certain cases. Obviously, the brand also tries to take advantage of this innovation in commercial terms, even registering the name of the innovation. Unfortunately, this type of argument is only valued by people who are duly informed and sensitive to environmental issues. Another very different case, but no less important, is the case of the steel manufacturer SSAB Swedish Steel, with factories in Sweden. While the Lulea plant is concerned about the environment with the energy consumed, seeking that it be supplied from renewable sources, trying to exclude fossil sources as much as possible, the Borlange plant is dedicated essentially to the rolling and coating of steel sheet. The Lulea plant supplies slabs to the Borlange plant, moving these slabs by rail. The steel rolling process is a natural consumer of water, which is why the company is located on the edge of a river. Part of the water is evaporated due to the temperature at which the slabs are taken to the rolling operation (around 1200°C), while another part is redirected back to the river. However, this water is contaminated with oxides resulting from the reaction of water oxygen with iron at high temperature. The company purifies the water used in the rolling process, announcing that the purity of the water returned

to the river after being used in the process is superior to the purity presented by the water at the time it is collected from the river.

Although these cases do not apply automation as a basis for installing more environmentally friendly processes, they are examples of the effort that companies need to make to contribute to a less polluted planet, less subject to extreme weather phenomena, and with less global warming. This effort only fails because it is very small, it is still far from being completely disseminated, and because the general consumer still has very little sensitivity to the environmental problems we are experiencing.

Automation can provide a remarkable contribution regarding energy savings, shortening cycle times, reducing the level of rejected parts that would be scrap, among many other cases that could be cited here. Automation can also promote the reuse of equipment put out of service but which, with a properly studied upgrade, could have a new life, being reused and preventing them from being recycled or deposited in landfills. Through many small actions, engineering and automation can make a remarkable contribution to an environmentally better planet.

Then, and more focused on the contributions that automation can provide to more environmentally friendly industrial processes, some case studies will be presented that correspond to works developed within the scope of master's dissertations at the Polytechnic of Porto, Portugal, resulting from collaborations between academia and industry, which have been implemented and which offer unequivocal benefits for the economic and environmental sustainability of companies.

10.2 Case studies

Below, some practical examples will be described of how knowledge and automation can positively contribute to reducing the ecological footprint induced by the industrial sector. This contribution can be made through actions aimed at saving energy, saving resources, or even reusing equipment considered out of service for certain purposes, but which can still be very useful for others. Thus, it is precisely with such a case that a short series of case studies analyzed in this chapter begins.

10.2.1 Using an out of service robot to build a new welding cell

The automotive industry is a large user of robots, from moving components during the different stages of a component stamping process, to painting. Moving from one model to another implies, in certain cases, the duplication of means, i.e., building a new assembly line, keeping the previous one active until the last moment. This is also related to the fact that adjustments are needed on new lines, which implies downtime that would lead to a production stoppage, which is unacceptable for most car builders. This implies that each new line is built with new robots and automatic systems, disregarding everything that already existed. Although there is a market for used equipment, since it has already been largely used, that equipment becomes unacceptable for a good part of used equipment market. However, the equipment is not obsolete regarding many other less demanding applications. This reality does not only occur in the plants of the main car builders, it also happens in small independent constructors and metalworking companies. The case described below is related to one of these situations, in which equipment still capable of providing a useful service is prevented from being sent for recycling or landfill.

This case study describes a use that was made of a robot previously taken out of service in another production line of tube frames for buses. Given that the production rate for buses is not the same to that of the passenger car industry, there is a good perception that this is an industry that could be the destination of previously used equipment (reuse). The bus frames industry starts with a chassis purchased from a supplier, which includes the front frame, engine and respective transmission and suspension systems, as well as the steering system. Starting from this base, a general frame is built, usually based on square or rectangular tubes, where the exterior panels and the entire interior set are then laid, from the floor, side and top panels, windows, seats, etc.

The work that now is intended to describe is based exactly on the bottom frame, which will link the front to the rear of the bus. This frame was positioned on a jig and welded manually, given that the initial intended production rate was low. With the number of orders rising and given that human labor is less prone to guarantee reproducibility and the required quality levels, the use of automation came to be considered, implying the design of a new welding cell. The frame of a bus is made up of several truss structures that are then joined, usually the lower part, with more than one level, the side parts, and the upper part, to which the front and rear parts are also joined. As the bottom part is the most difficult frame to weld (Fig. 10.1), due to their area and the deformations to which they can be subject, this was the by-products chosen for the work, given that if it is possible to meet the requirements in welding these frames, it would also be possible to carry out all the rest. First l, the necessary work area had to be established, in order to analyze whether the robot's working range would be sufficient. It was established that the welding would be carried out in two stages: a first dripping stage for the relative positioning of the different components with respect to each other, and a subsequent stage of consolidation welding. Given that the tubes are relatively thin, care should be taken with welding to minimize localized heating and, consequently, deformations and residual stresses that are difficult to reverse. The area requirements to be covered by the robot were established as 7500 mm long by 2500 mm wide. The robot available to serve as the basis for this work was a Fanuc Arc Mate 100i-B, previously used in another project. As the maximum available robot working distance is 1373 mm, the jig will need to be able to rotate around two points, joined by an axis parallel to the front line of the robot. However, even this option is not enough for the robot to reach 7500 mm away. Thus, it was also necessary to provide the system with a sliding guide for the robot, allowing it to move parallel to the side of the jig and the frame to be built.

To facilitate the rotation of the jig, the drive motors are equipped with synchronized vertical movement in each of the columns (Fig. 10.2, left). Thus, for welding the jig is placed in the lowest position and when it is necessary to rotate it to weld on the opposite side or in the vertical position, the jig is raised, and the simultaneous rotation of the two support points is activated. Regarding the robot's access to the points where welding is required, the 2500 mm width of the frame is not a problem, as when the jig rotates, the effective distance from the surface of the

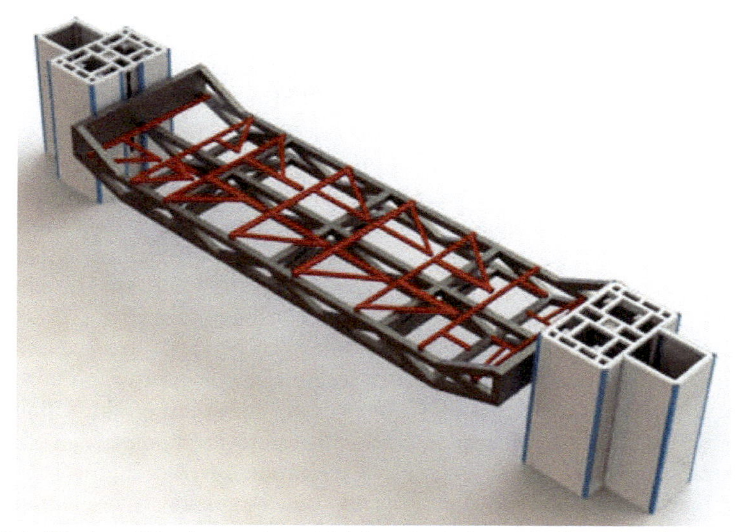

Figure 10.1. 3D view of the frame to be welded (in red color) over the jig (gray color), and rotational points at the ends of the jig.

Figure 10.2. Column holders for the jig rotation points (left); Table with the 6th axis, parallel to the axis of the jig rotation (right).

frame integrated into the jig is less than half that distance, i.e., smaller than 1250 mm. Given that, as previously mentioned, the robot's radius of action is 1373 mm, distance is no longer a problem. However, as one is talking about radius, the problem would only cease to exist if the robot is able to move along the jig, a situation that had to be implemented through a system of linear guides and an encoder. Thus, in addition to the five typical axes of a robot with this morphology, the system is now equipped with a sixth axis, which allows the robot to move linearly along a line parallel to the axis of rotation of the jig (Fig. 10.2, right). With this solution, considering the jig in a vertical position and the robot arm positioned vertically to its axis of movement, as well as the axis of rotation of the jig, the maximum distance it would need to reach would be, hypothetically, 1250 mm, which is perfectly achievable by the robot, given its range of action. Furthermore, it is possible to rotate the jig to the position that is most favorable at any time, further shortening the distance between the robot and the frame to be welded. Basically, this axis works as a 7th axis, not of the robot itself, but of the positioner, which in this case is the jig that supports the components, which will give rise to the frame.

Once implemented, and still during the programming and testing phase, in addition to positioning using dripping, the frames also began to be fully welded onto the jig by the robotic system. The accessibility to each point of the structure was analyzed, minimizing the jig material at the points where the welding torch would be more critical, thus facilitating the welding of the entire frame. The thermal input was also studied to avoid exaggerated punctual heating, which would translate into high residual stresses and, consequently, deformations. After an initial simulation period, and then with the robot already installed, all the joint programming of the different devices that make up the system was carried out, verifying that after the final adjustments, the total welding of the frames was perfectly feasible.

In environmental terms, this project essentially served two purposes: (a) it allowed the reuse of a robot that was sentenced for recycling or landfilling; (b) the reliability of the process generates a practically irrelevant number of defects, also saving resources in this way. It is true that the process will consume some energy in feeding and moving the robot and the jig, but, in general terms, this is largely compensated by the advantages referred above.

In addition to the environmental aspects, the project aimed at increasing the production capacity of the bus frame production cell. Effectively, production time in the manual mode (human labor) was 35 min per structure, falling to just 17 min per structure with the implementation of the robotic cell. As for the cost of implementing the project, and considering the reuse of the robot, which was previously amortized by the project where it had been used, was fixed at € 35,500, a cost that concerns the robot's linear movement table and the rotation supports from the jig. The cost of the jig was not considered in its entirety, as it was also reused from the manual process, with only the upgrade/adaptation cost being accounted for. Considering the time saved in each structure, a study was carried out regarding the time needed to obtain a return on investment, obtaining a payback period of 21 mon. This proves its economic sustainability, beyond the environmental gains. However, social sustainability has also been improved. The welder was taken to training courses in welding inspection, which, despite decreasing in time per unit produced because the quality assurance, it also became necessary to reinforce human resources due to the increase in the number of units produced. As a result, the welder is no longer exposed to fumes, radiation, and less ergonomic welding positions, which could lead to occupational illnesses. It can be observed that this project substantially improved the sustainability of the process in all its aspects [1].

Manual welding is more subject to generate defects. To establish a comparison, 20 manually welded structures were analyzed, with an average of two defects per welded frame being observed. After the implementation of the project, and after the necessary adjustments to put such equipment into operation, i.e., after the process has been stabilized, 20 frames were also analyzed, and no defects were detected. This observation shows the reliability presented by the robotic system in relation to manual welding, which is an aspect to be highlighted, since it will reduce the control effort from a 100% analysis to a sample analysis. Moreover, there is also a saving in terms of resources, due to less scrap generation.

10.2.2 Using an out of service robot to build a new assembling cell

This case study describes the use of an out-of-service robot to build a parts assembly cell with some components in freshly injected components. The problem intended to be solved is directly related to the optimization of a component inspection and assembly process for the automotive industry. The case study was essentially centered on a plastic cover for the B-pillar of a motor vehicle. The part in question is injected in PP (Polypropylene) at high pressure into the mold and is removed by mechanical extraction. These parts fit into the car's body using two clips that are inserted by operators after an injection. The area where the clips are housed is called clip holders.

The mold installed on the production line is a plastic injection mold with mechanical extraction. The component under study exists on both sides of the vehicle in which it will be assembled (right B-pillar and left B-pillar), so two parts are produced in each injection cycle, one for each side of the vehicle. Since it is common that the existence of channels in plastic injection through which the plastic is injected into the mold, this does not occur in the present case study, in which the plastic is injected directly into the cavity of the part in the mold, so it is not necessary to remove the parts sprue, which normally comes out connected to the finished injected part. The injection molding machine usually used in the production of this set of parts has a capacity of 800 tons of closing force, and production parameters such as injection temperature, pressure, cycle time, etc., were properly optimized for production of these parts. The production line responsible for the injection and finishing of these parts consisted of an injection machine, a three-axis robot responsible for extracting the injected parts (Sepro) and placing them on the adjacent conveyor belt, two operators responsible for the first visual checks, and two peripheral equipment whose function is to insert the clips and carry out the necessary validations. In Fig. 10.3, a sketch of the current layout and the flow of parts from the mold to the initial storage station, which will then be stored at the dispatch station, is presented.

Initially, the quality control of the parts was carried out in the peripheral equipment schematically represented in Fig. 10.3, that is, the parts were injected into the mold, then they were extracted onto a conveyor belt that served to transfer these parts to the operators who were between the conveyor belt and the peripheral control stations. These operators proceeded to assemble the clips and deliver the parts to the peripheral control equipment, where certain parameters of the parts were checked, such as the possible existence of incomplete parts, a general visual inspection of the parts, the correct insertion of the springs, and verification of the version of the part produced (with or without the "air-bag"). In case the previously defined quality parameters are confirmed, it becomes necessary to perform another very important task on the validated parts, using the same peripheral equipment. The parts are marked with a notch to improve the parts traceability and ensure that the parts have passed through the equipment and have been duly validated. Thus, it is mandatory that there are two pick points for the part, one to verify its passage through the equipment, and another to verify the respective validation. The chipping

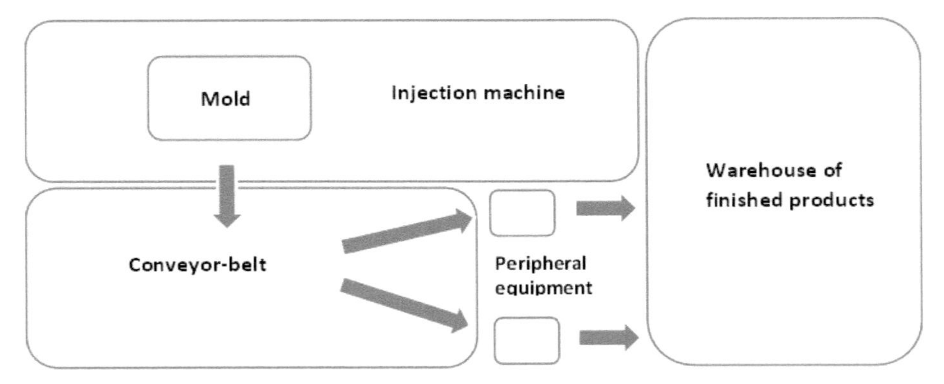

Figure 10.3. Schematic diagram of the workflow at the station under study.

is carried out by a pneumatic actuator, which leaves a slight mark on the part, usually in what is known as the technical area. After a study of the sequence of operations to be carried out, as well as the time related to each operation, it was found that the post responsible for the bottleneck in the process flow is the injection. Thus, it became necessary to develop the process so as to make it more expeditious, bypassing the bottleneck identified. Given that it is not possible to speed up the injection process, under penalty of producing defective parts, it is necessary to rethink all other tasks, with a view to improving its competitiveness.

Furthermore, it became mandatory to modify the workflow to improve the efficiency of the production line. It was understood as a priority to distribute the workflow after the parts were extracted from the injection molding machine, and to automate the operations that were being carried out by the operators. Effectively, if the general cycle time is conditioned by the injection molding cycle, the only way to decrease the costs associated with the process would be to reduce the allocation of labor. The automation of processes would imply the introduction of a robotic cell between the injection molding machine and the conveyor belt, allowing to anticipate some operations that were concentrated in peripheral equipment. A new configuration of the production cycle then emerged, which is shown in Fig. 10.4, based on Fig. 10.3.

In the new workflow, the stages of the earlier inspection—clip assembly and marking—are now carried out by the robotic cell, and now only one operator is required, who, in addition to inspection, also starts to pack the parts and deliver them to the warehouse in a ready-to-ship condition for shipment to the end customer. For this robotic cell, some requirements were defined: (a) it would have to incorporate a robot, which would collect clips from a feeder and insert them into the injected parts; (b) as the parts are placed by the Sepro robot which is suspended, the cell should receive the parts dropped by that robot and insert the clips from the bottom; (c) a feeder system capable of supplying the clips necessary for the correct operation of the new robot is required; (d) the base of the cell acts as a jig for housing the parts left by the Sepro robot; (e) the base should be flexible enough to easily adapt to possible modifications of the parts under possible further requests of the customer; (f) after the Sepro robot drops the parts into the cell, they must be subjected to a fixation

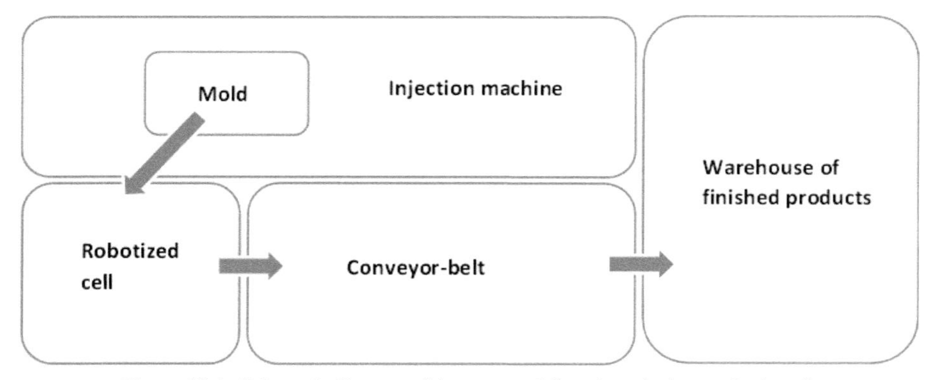

Figure 10.4. Schematic diagram of the new workflow, introducing a robotic cell.

that allows the insertion of the clips by the new robot, without the parts moving; (g) the cell must be equipped with appropriate sensors and an artificial vision camera to detect defects such as 'poor filling' or flow streaks on the surface of the parts. In addition to all these requirements, it became essential that the project reused as many components as possible, thus contributing to economic and environmental sustainability. Thus, and taking advantage of cells designed earlier for other purposes, an existing frame with adequate dimensions was reused, to avoid dimensioning and building a new one. Given that the necessary robot does not need to be technologically very advanced, as it aims to collect clips from a feeding system and place them in the respective parts, an out of service robot was reused, coming from a discontinued project. With this option, it was avoided that an equipment and respective controller were to be recycled, when they could still be useful in the intended functions. Part of the electrical panel was used, in terms of components, thus also making it possible to care for the environment and save on investment. Among the cylinders that had been taken out of service in the meantime, the most suitable ones were selected in terms of load capacity and stroke to be reused, with a view to fixing the parts to the template. Only the jig, the clip feeding system and the gripper for placing the clips in the injected parts needed to be designed according to the new needs. With these reuses, it was possible to greatly reduce the investment in the new robotic cell, giving a new useful life to the robot, to several mechanical components (frame) and to some electrical, pneumatic, and mechanical components (automation, fastening systems and cylinders). Thus, the economic and environmental aspects were taken care of jointly.

The robotic cell now performs various control operations using sensors suitable for the purpose, namely by reading the presence of material in places where the presence of defects was traditionally detected in the injection, analysis of possible flow rays in the part through artificial vision, as well as the presence or lack of clips in the appropriate places. Once this verification has been carried out, the same cell is able to mark a hexagon on the parts, ensuring that they are alright and have been controlled. After this, the parts are sent to the conveyor belt, where the operator makes a final visual check, packs the parts, and sends them to the warehouse already in the conditions in which they should go to the final customer. Some of these systems can be seen in Fig. 10.5.

A. Base with protective frame

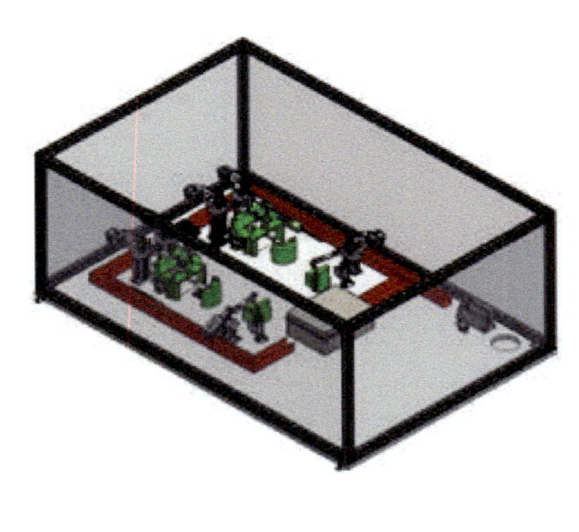

B. Clamping system for workpieces

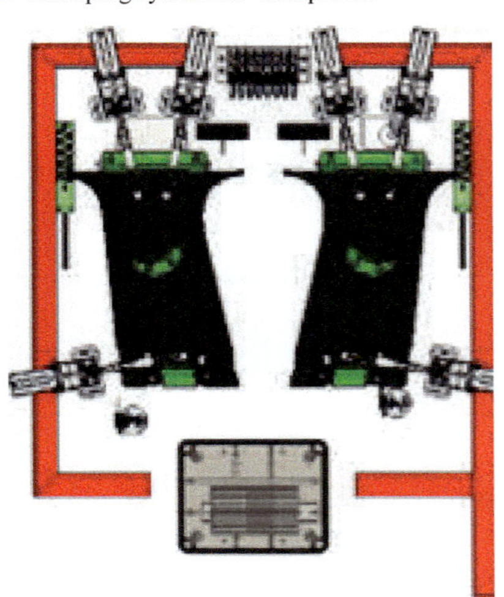

C. Artificial vision system for control

D. Clip insertion robot

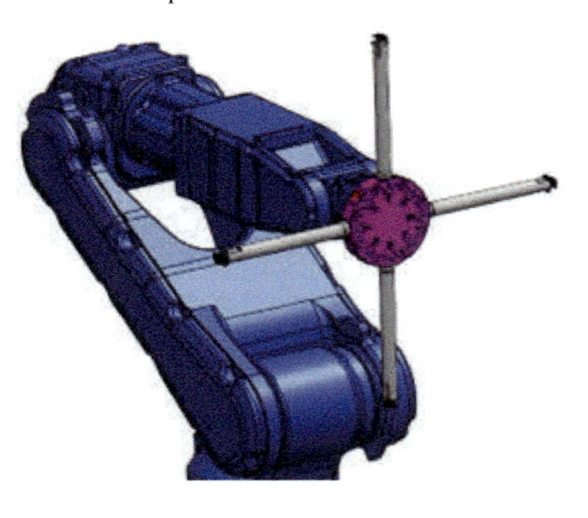

Figure 10.5. Main devices making-up the new robotic cell, being largely reused or recovered (base and robot).

After the new robotic cell has been designed, and after its validation in industrial terms, it was implemented. Given that most of the devices and materials applied had been used before and were duly amortized by the projects where they had been inserted, only four components contributed significantly to the costs of the new cell: (a) Clip feeding system – € 4,500; (b) Gripper and other accessories – € 2,000; (c) Aluminum base – € 150; (d) Labor and engineering – € 10,000. Thus, the total cost was € 16,500, an amount that can be easily amortized by the gains in productivity achieved. These gains translate into the continuous reduction of one operator (estimated at around € 1,300/work shift), a more continuous workflow, without waste of time, reduction of non-compliance costs due to more accurate control parts, and reduction of possible failures due to human fatigue. On the other hand,

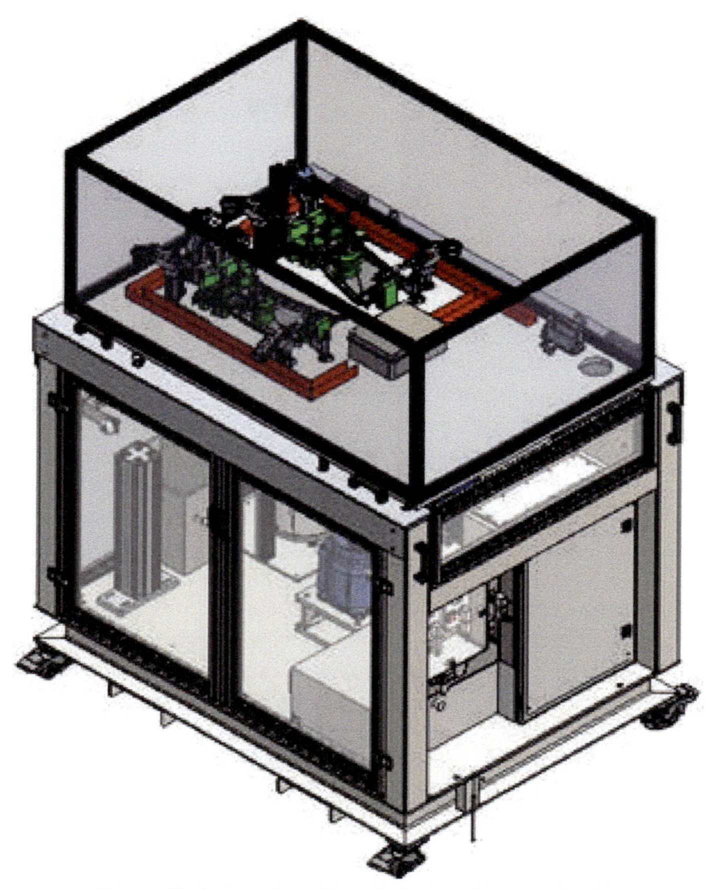

Figure 10.6. Overview of the 3D design of the robotic cell.

the following equipment and devices were reused: robot and controller, electrical panel and respective components, cell protection frame, aluminum frame, and all pneumatic material, including the valve block and 33 pneumatic accessories. The robotic cell design can be seen in Fig. 10.6.

Thus, it was proven that companies can have significant gains in competitiveness if they properly analyze their processes, and engineering puts its knowledge into practice. Taking advantage of numerous components and equipment that are no longer needed in end-of-life projects, it is only necessary to develop solutions that, considering the existing material, can produce solutions capable of solving competitiveness problems that could arise, due to greater efficiency of the competitors, or competition induced by markets where labor costs are still very low. Solutions like that can motivate many other companies to follow the same way, improving their economic and environmental sustainability regarding the guidelines described here [2].

10.2.3 Saving energy by automating an over-injection discharge process

The automotive industry consumes numerous components, and Bowden cables are one of those components, indispensable for mechanically opening doors, among others. The manufacture of these cables involves a large number of operations, which will be described in more detail in the following case study. Among these various operations, there is a plastic injection, which is carried out after a wire-rope has undergone the 1st injection of Zamak (zinc/aluminum/ magnesium/copper alloy) in its extremity. Bowden cables are also made of a spiral tube covered by extrusion of plastic over a laminated steel wire and spirally wound. This gives it very interesting elastic properties, and adequate mechanical resistance to let the wire-rope flow inside the car doors, floor, and so on, often following complex layouts. The injected part is called grommet and needs to be injected onto the spiral tube. This injection is carried out in a group of parts that are assembled in a specific mold, normally with eight cavities. This grommet is normally responsible for fixing the Bowden cable to the separation between the wet and dry zones of car doors, being solidary with the coated spiral tube in a determined position and functioning by fitting in this separation. The still incomplete product (by-product) is very flexible but needs to be carefully assembled in the respective mold where it will undergo this plastic injection. Due to this flexibility, automating the operation of incorporating the spiral tubes into the mold will have to be explored later, focusing now on the extraction operation. In fact, after the injection, the by-product is correctly positioned and can be automatically picked-up. Each injection machine works with a mold and a half. Effectively, there is only one upper half-mold for two lower half-molds. This allows to optimize the work during the injection and cooling time. Each operator initially worked with a single piece of equipment, taking advantage of the injection and cooling cycle time to remove the already eight injected parts from the half-mold that opened, clean it inside and place new spiral tubes for the next injection cycle.

The main challenge posed in this case study was related to the high energy consumption of the equipment used, which was not minimally proportional to the production flow achieved. This situation made clear a lack of efficiency, which was high enough to justify an intervention. In fact, the energy consumption for each full working day considering three shifts was initially 543.26 kWh/d for a target of 8,756 injected parts per day, using a J. B. Fiser® injection machine provided with an eight-cavity mold. However, there are currently injection machines in the market using much less energy to produce the equivalent work. The solution adopted was a BabyPlast® injection machine, which consumes only 54,82 kWh/day in the same working conditions, but only permits to use a four-cavity mold, allowing a production of 4,762 injected parts per day and machine, without automated extraction, regarding the forecasted cycle time (see Table 10.1). Thus, this solution is not enough to reach the production pace required for the project. If one single machine is not enough to produce the quantity of injected parts needed, the use of two of those machines need to be considered, because even using two BabyPlast® machines, the energy savings would be very significant (about 80%). However, the investment needs to be studied. This solution requires two operators in charge, because the operation cycle initially

Table 10.1. Details about the operations in injecting coated spirals used in Bowden cables.

		Processing time (s)	
		1 x J. B. Fiser	**2 x BabyPlast**
Operation	**Operation details**	**Initial**	**Final**
Positioning conduits	Inserting 8 conduits on the mold in the proper grooves	13	14
Aligning conduits	Operator adjusts the flexible conduits into the grooves, in extension and position	14	14
Start injection	Operator pushes the button to start the process, closing the mold	5	6
Injection	Molten plastic flows on the mold channels, filling the corresponding cavities	30	10
Mold opening	Operator pushes the button that generates the mold opening process and moves the lower half molds between the two possible positions	5	10
Conduits extraction	Operator removes the 8 injected parts and put them into a box, removing the scrap as well	7	14
Average time for 8 injected parts		74	68
Average time for 8 injected parts with automatic extraction			44
Number of injected parts per day (22.5 hr real working time)		8,756	9,529
Number of injected parts per day with automatic extraction		----	14,727
Saving in terms of energy			~ 80%
Increase in terms of productivity			~ 68%

undergone requires one operator by machine to charge the coated spirals (also called as conduits) to the mold and collect the spirals already provided with the grommet. Thus, automation would be the only solution to allow to pursue with this project. The idea behind the use of BabyPlast® injection machines was to optimize the production cycle, putting each operator to work with two BabyPlast® machines. However, because the loading operation is complex due to the low stiffness presented by the coated spirals, the efforts need to be focused on the mold unloading operation, saving time to unload the mold through the introduction of an automatic unloading system. To ensure the general cycle time in each of the cases, as well as the production volume, a diagnosis was made for the initial situation, and another for the situation that was intended to develop. The study is described in Table 10.1.

Considering the use of two BabyPlast® machines and the automatic extraction system to be developed, the times highlighted in red color in Table 10.1 could be cut, increasing the productivity. In fact, if the worker is operating two machines, some of the work is made in parallel and can be removed from the total production cycle time. In a rough approach, it can be considered that the sum of the mold opening time and the injected parts extraction time is almost equivalent of the previous operations (positioning conduits, aligning conduits, start injection and injection times). The difference is the time needed for the worker to rotate into the workstation between injection machines.

Thus, with the cycle times balanced, it was time to develop a solution for extracting the injected parts. Penne et al. [3] promoted an indepth study on the movements needed to insert the conduits and extract the injected parts from the mold and drop them into an adjacent container, allowing their easy conveying to the next workstation where the Bowden cables' manufacturing process will pursue. First, a diagram of the movements needed to be performed by the operator was drawn, as depicted in Fig. 10.7 (from position 1 to 2, from 2 to 3, and so on, returning at the end to position 1). The operator just needs to feed each lower half-mold with four conduits and start the injection cycle, because the remaining actions will be performed by the injection machine and by the automatic extraction system to be developed. Thus, the principle of the system was set-up, being necessary to design it based on the vertical and horizontal movements the system needs to perform, extracting the over-injected parts and depositing them into the corresponding containers. The layout of the automatic system developed, and corresponding workflow can be seen in Fig. 10.8, as well.

As there are two lower half-molds, the automatic system will have to be symmetrical, i.e., allow movement from one side to the other of the injection molding machine. Obviously, the movement systems will have to be supported by a common robust frame, which ensures the necessary movement accuracy. The system needs a vertical movement to extract the injected parts from inside the mold, and a lateral movement to take them to the position where the containers are located, and drop the parts previously removed from the mold. For this horizontal and vertical displacements of the gripper system, a lever was selected (Fig. 10.7), where the system that collects the injected parts from inside the mold will slide. To collect the

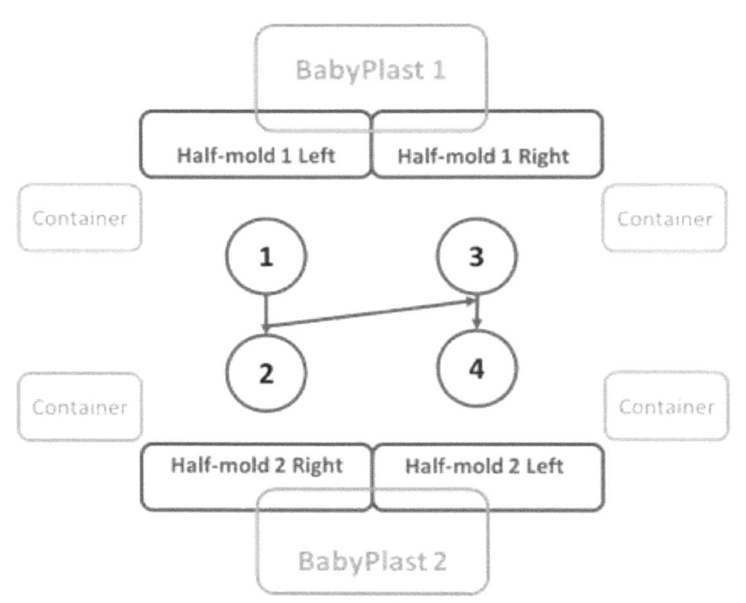

Figure 10.7. Workflow of the operator with two BabyPlast® injection machines, alternating positions 1, 2, 3 and 4.

injected parts from inside the mold, a set of grippers is needed, i.e., as many as the parts to be collected (four) on each side of the injection molding machine. The mold needs to have its own cavities so that the end of the grippers can exert the necessary force to extract the parts from the mold. The movements to be carried out will have to be synchronized, but they are relatively simple to program, since they are linear, and there are no accentuated problems of possible collisions. Thus, some details of the systems pointed out in Fig. 10.8 can now be observed in Fig. 10.9.

It should be noted that the frame needed to be dimensioned for the expected efforts, and the pneumatic cylinders needed to be selected so as to comply with the planned route and the previously defined cycle time.

The final solution allowed obtaining an energy reduction from 543.26 kWh/d to 109.64 kWh/d, corresponding to a percentile difference of 79.2%. The cycle described by the handler takes 6 sec in practical terms. In addition, it was also possible to achieve a reduction in plastic waste from 480 kg/yr to 61.44 kg/yr, which represents a saving of 87.2%. The concept presents high flexibility, being adaptable to other Bowden cables models. According to a study carried out for return on the investment, it was found that after 66 d of operation with three shifts, the return on

Figure 10.8. Concept of automatic system to extract the over-injected parts from the mold and deposit them into the containers.

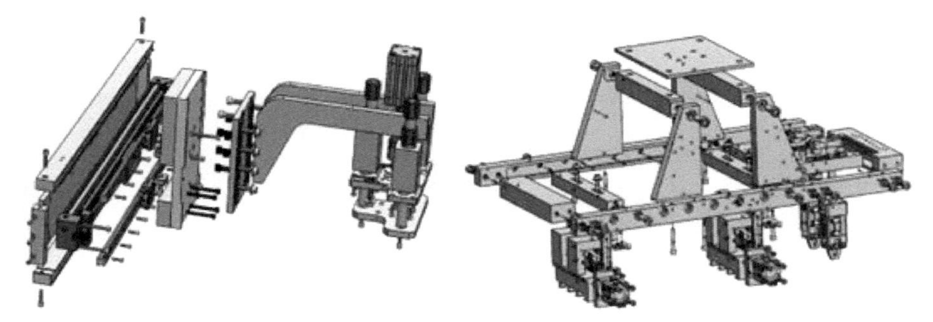

Figure 10.9. Details of the lever system which allows the horizontal displacement of the gripper, as well as the vertical movement of the gripper (left), and gripper system, which will be assembled in the lever system (right).

investment would already be completely achieved. This case study clearly shows how automation can contribute in different ways to increasing sustainability, through solutions with a high return on investment.

10.2.4 Designing a micro-injection machine aiming to save energy

Still in the field of the polymer injection, the commonly used injection machines have their own capabilities for relatively large parts, hence bunches of parts are produced in each injection, taking advantage of the maximum capacity of the machines. However, there are also components, particularly for the automotive industry, although not exclusively, which require very small amounts of injected polymer, and where it is a waste to expend immense energy, when the volume of plastic to be injected is so small. In addition, the volume occupied by these machines is also an obstacle to their being inserted in automatic production systems, which integrate different processes.

As previously referred to in other case study, Bowden cables are automotive components that activate certain devices from inside the car, such as the doors, hood, trunk lid, fuel tank lid, window raising and lowering, parking brake. mechanical hand, etc. With a view to reducing the noise at the ends of these cables when the car is moving, the most reputable brands promote a polymer coating of the terminals injected in Zamak, which normally weigh between 5 grams and 15 grams. This coating can be made of polyethylene, polyamide (with or without the addition of short fiberglass reinforcements), or polyacetal. Moreover, these polymers are also used to add other components by injection to the Bowden cables, as shown in Fig. 10.10.

Bowden cables have undergone several improvements in their production process, moving from batch production to series production. Due to the high number of processes involved in the manufacture of Bowden cables (wire rope cutting process, stripping, 1st Zamak injection, spiral tube assembly, foam tube assembly, plastic injection, punching, engraving, 2nd Zamak injection), each batch of cables was subjected to one operation at a time, then moving on to the next post. Successive efforts by some researchers [4-6] led to a new approach in the manufacture of these components for automobiles. Thus, the process became continuous, with the wire-rope being cut and all the following operations being added in the same piece of equipment, with the wire-rope meanwhile undergoing additions, until it becomes the Bowden cable that is sold to the customer. This requires a step-by-step system that allows adding new features to the Bowden cable. But, for that, the necessary peripheral equipment will have to be coupled to the main equipment. The larger this peripheral equipment, the larger the Bowden cable production core equipment should be. Hence, the efforts to reduce peripheral equipment, so that the step between different operations can be as reduced as possible, without conditioning future maintenance operations. Integrating large injection machines into Bowden cables production equipment is something unthinkable, due to the proportions of the components and equipment involved. Furthermore, the energy consumed would be extremely high. Previously, the injection process allowed the inclusion of four to eight cables at a time, but the injection capacity of the equipment is usually

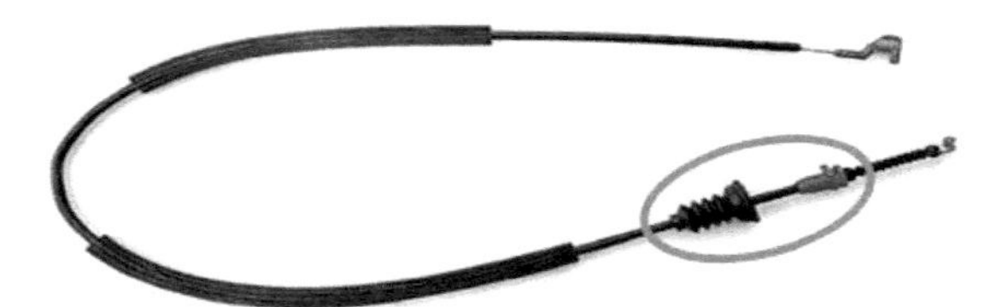

Figure 10.10. Overview of a Bowden cable, with polymeric injected components highlighted.

much higher. As a result, energy consumption was also higher, which is not a very convenient situation for the environment.

Vieira et al. [7], at the request of a company that produces Bowden cables and manufactures its own equipment, developed a thermoplastic microinjector for the injection of small components in Bowden cable, with a view to minimizing energy consumption, maintaining, or increasing the cycle time obtained until the moment, which is compatible with the cycle time of the automatic production machines for Bowden cables, thus increasing the competitiveness of the product. The injection machines normally used to carry out injection operations on cables are the JB Fiser, model VMCC 80/200 (Barcelona, Spain), consuming a lot of energy and being clearly disproportionate to the volume of material to be injected, even when several parts are grouped together. With a view to a concept test, a Babyplast® equipment (Molteno, Italy) was tested, which was already better suited to the volume of material to be injected. As a concept, the equipment worked, but the intention was to develop a specific equipment that could be easily incorporated into the automatic machines under development.

However, for this development, it became necessary to define the clamping force of the mold, which depends on the volume of the parts to be produced. Thus, to start the project, it was agreed that the part would have a volume of 1.1483 cm^3, would have a revolution shape, could be injected in High-Density Polyethilene (HDPE), Low-Density Polyethilene (LDPE), Polyamide (PA666) or polyoxymethylene/polyacetal (POM), and the maximum cycle time for a part with these dimensions was defined as 6 sec. The main motivations were: (a) saving cost and the environment; (b) integrate all functions in the same console of the equipment; (c) have a better control about all the peripheral equipment, improving maintenance operations.

The initial injection molding machine (JB Fiser) required the use of two molds, to save cycle time: while one mold was receiving molten polymer and cooling down, the operator was working on the other, removing the finished injected part and preparing the mold for a new injection. In fact, there is a lateral movement of the table that supports the molds in the injection machine, and promotes the alternation of work in one or the other, according to the operator's order, and assuming the injection cycle is finished. At each cycle, four to eight cable ends were injected, forcing the operator to correctly position them longitudinally and laterally.

To validate the concept and study the cycle time, a test was initially carried out with Babyplast® equipment, which did not meet some of the project requirements, such as allowing the control to be integrated with the main equipment. However, some of the purposes for using the equipment, such as energy saving, were already partially or fully fulfilled by this equipment. Considering the cycle time, molds with four cavities were made, given that the Bowden cables will be moved two by two by the equipment's step-by-step system. Images from this initial experiment can be analyzed in Fig. 10.11.

Based on this experience, some requirements were defined to be met by the future micro-injector, such as: (a) being compact and easy to integrate into the automatic production equipment for Bowden cables production; (b) possibility of processing different materials and dimensions of parts, within a previously defined range; (c) cycle time not exceeding 6 sec ; (d) inclusion of a heated injection nozzle to prevent plastic incrustations on the mold; (f) reduce energy consumption, from current 30 kW to a maximum of 10 kW. The design of the new equipment included the selection of sensors and actuators to be integrated by the PLC (Programmable Logic Controller) of the main equipment, and sizing of the hydraulic system serving the micro-injection machine. Since there is a common water-cooling circuit, the injector only needed to inspect the water inlet and outlet connections.

The new equipment is essentially constituted by three systems (Fig. 10.12): (a) the compression unit, which ensures the movement of the mold when necessary, and the non-opening during the injection and cooling period; (b) the injection unit; (c) the hydraulic unit.

Among the materials likely to be injected into the equipment to be designed in the future, the one with the most limiting characteristics is PA66: melting temperature of 260°C and injection pressure between 1000 bar and 1600 bar. Thus, the equipment was sized to work with these properties. It should also be observed if PA66, among the considered polymers, is the most hygroscopic. Using the Wubken equation, the time required for cooling was estimated considering the various polymers to be considered, with polyacetal being the one that most severely limits the cycle time, with 35 sec. Considering the established cycle time (6 sec) and the maximum cooling time, 35 sec, the number of parts to be injected per cycle was calculated so as to maximize the efficiency: seven parts (5.5 sec/part). Based on this value, the total volume was calculated, 8.04 cm^3, and the necessary clamping force: 75 kN. Subsequently, it was necessary to dimension the frame of the pressing system as a way to withstand the maximum expected forces.

The estimated cost for the micro-injector is €23,025, which includes raw materials, standard components, specific components, manufacturing, and assembly. Considering three shifts/d of work and 235 d /yr of work, for the considered cycle time, around 3.3 million parts will be produced. The energy consumed by JB Fiser is 124.50 MW/yr, while the new equipment consumes only 12.60 MW/yr, i.e., there is a real energy saving of 89.9%. Considering an energy cost of 0.1481 €/kWh, energy savings for the same quantity of parts produced, and considering the costs of equipment and molds, it appears that the return on investment is achieved in 1.2 yr. The investment is usually recommended under these conditions. In addition to these factors, which are decisive for environmental and economic sustainability, this equipment provides the possibility of being assembled as a peripheral to the Bowden cables production main equipment, which is also an extremely important factor.

Although investments are almost always motivated by economic reasons, in this case, the environmental benefits are equally or more important than the financial gains.

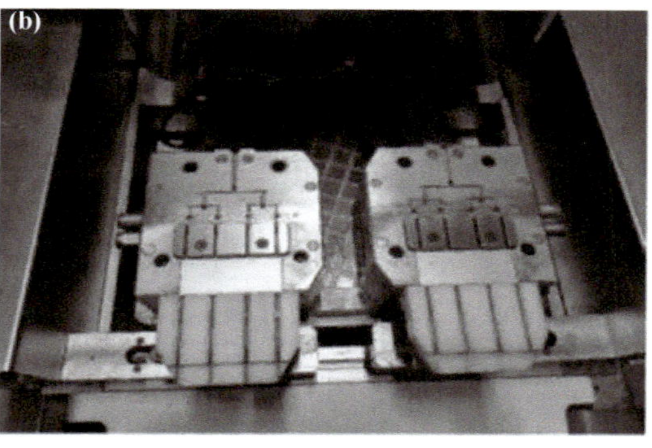

Figure 10.11. (a) Babyplast® injection machine and (b) corresponding molds.

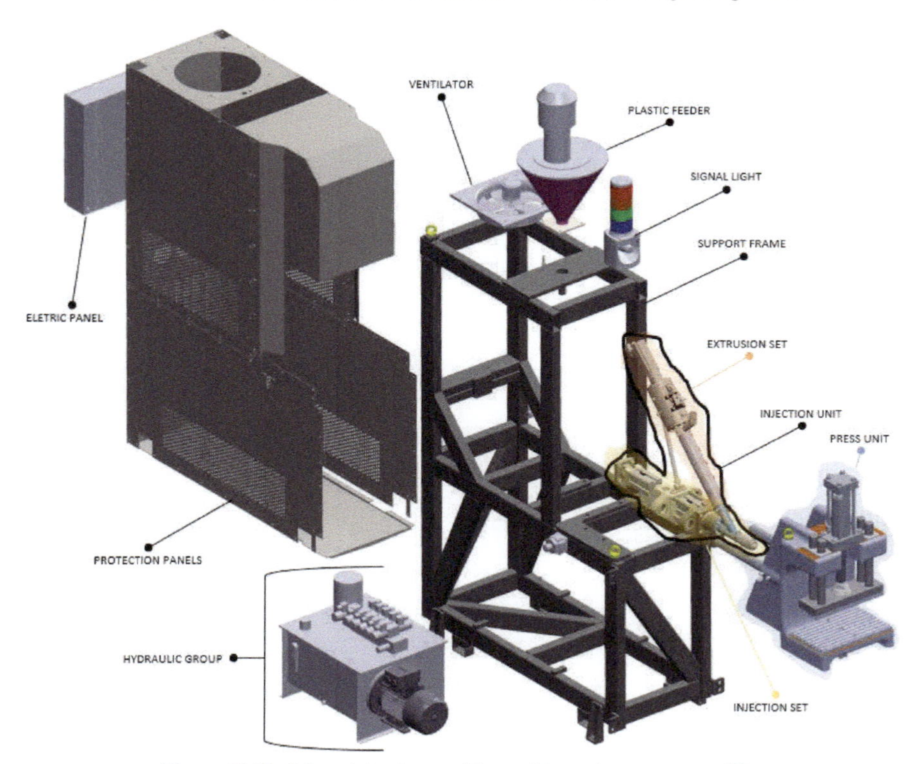

Figure 10.12. Micro-injection machine and its main components [7].

Reducing the power consumption by 89.9% of the energy initially consumed is an excellent result. Thus, this is also a great example of how engineering and automation can significantly improve environmental sustainability.

References

[1] Castro, A. F., Silva, M. F. and Silva, F. J. G. (2017). Designing a robotic welding cell for bus body frame using a sustainable way. Procedia Manufacturing, 11: 207–214. doi: 10.1016/j.promfg.2017.07.225.
[2] Costa, R., Sousa, V., Silva, F. J. G., Campilho, R., Sá, J. C., Pinto, A., et al. (2022). Increasing the Sustainability of manufacturing processes in plastic injection: recovering out-of-service robots to eliminate manual assembly operations. Sustainability, 14: 12300. doi: 10.3390/su141912300.
[3] Penne, R. Silva, F. J. G., Campilho, R. D. S. G., Santos, G., Sousa, V. F. C., Ferreira, L. P., et al. (2022). A new approach to increase the environmental sustainability of the discharging process in the over-injection of conduits for bowden cables using automation, Proceedings of the Institution of Mechanical Engineers, Part C: Journal of Mechanical Engineering Science, 236(16): 8823–8833. doi: 10.1177/09544062221087547.
[4] Moreira, B. M. D. N., Gouveia, R. M., Silva, F. J. G. and Campilho, R. D. S. G. (2017). A novel concept of production and assembly processes integration. Procedia Manufacturing, 11: 1385–1395. doi: 10.1016/j.promfg.2017.07.268.
[5] Sousa, V. F. C., Silva, F. J. G., Campilho, R. D. S. G., Pinto, A. G., Ferreira, L. P. and Martins, N. (2022). Developing a Novel Fully Automated Concept to Produce Bowden Cables for the Automotive Industry, 10: 290. doi: 10.3390/machines10050290.
[6] Vieira, D., Silva, F. J. G., Campilho, R. D. S. G., Sousa, V. F. C., Ferreira, L. P., Sá, J. C., et al. (2022). Automating equipment towards industry 4.0: A new concept for a transfer system of lengthy and low-stiffness products for automobiles. Journal of Testing and Evaluation, 50(5): 2310–2325. doi: 10.1520/JTE20210721.
[7] Vieira, A. L. N., Campilho, R. D. S. G., Silva, F. J. G. and Ferreira, L. P. (2021). Increasing the environmental sustainability of an over-injection line for the automotive component industry. Sustainability, 13: 12692. doi: 10.3390/su132212692.

Index

3D printing 116, 129, 130, 132

A

Additive manufacturing 116–119, 121–134
Applications 118, 119, 127, 130–132
automation 36–46, 48, 50, 52–60, 62, 64–68, 74, 78, 80–83, 87, 91, 171, 182–185, 187–193, 195–199, 202–204, 208, 209, 213, 216, 219

B

Being greener 11

C

Chemical Vapor Deposition 149, 156
Coatings 140, 142, 144–158
construction industry 171, 172
Consumerism 1, 4
Corrosion resistance 147, 149, 150

E

Ecological footprint 36
Effluents 142, 155, 158
Energy management 190–193
Energy saving 188–190, 195
Environment 95–103, 106, 107, 111–113, 140–143, 145, 147, 149, 155, 158, 182, 183, 186–188, 191–193, 195, 196, 198
environmental concerns 171, 176, 178
Environmental education 4, 5, 12
Environmental sustainability 80, 82, 84–90, 182, 184, 187, 188, 197, 198

F

Family life 182

G

Greener industry 62
Greener processes 202
Greener technologies 163

H

Home 183–195, 199
Home automation 183, 187–190, 192, 193, 195

I

Improving processes 113
Industry 74–76, 82, 87, 91
Inequalities 19, 22

L

Lean green 87, 90, 91

M

Manufacturing 74, 77–80, 83, 87–89
Materials 118–123, 125–133

P

Physical Vapor Deposition 146, 150, 154
Political actions 10
Population growth 1, 4
Process industry 100–102
pulp and paper industry 175–177

R

Reconfigurable manufacturing systems 80, 83
Recycling 116, 120, 122, 127, 130–134
Reuse 203, 204, 206
Robotics 78, 80, 87, 91, 206, 208–211
Robots 37, 40, 51, 52, 57, 59–61, 62, 67, 68

S

Safety 89, 91
Saving energy 203, 212
Saving resources 38, 39, 48, 50, 57, 58, 62, 67, 103, 118, 121, 203, 206
Sensors and actuators 185
Smart cities 198
Smart manufacturing 95–101, 103, 104, 107, 108
Society 17, 18, 20, 22, 24, 28, 33
steel making industry 169
Sustainability 98, 99, 105–107, 111–113, 116, 119, 121–129, 131
Sustainable development 17, 27, 28, 31, 32

T

textile industry 177–179
Thin coatings 153, 158

W

Wastes 133
Wear resistance 144, 155